I0041782

© 2024, Digital Craft Publishing 2024

All rights reserved. Apart from any fair dealing for the purposes of private study, research, criticism, or review, as permitted under the Copyright Act 1968, no part of this publication may be reproduced, stored in a retrieval system, or transmitted in any form or by any means, electronic, mechanical, photocopying, recording, or otherwise, without prior written permission from the copyright owner.

Any unauthorised commercial use of this publication or its contents is strictly prohibited. The moral rights of the author have been asserted.

First published in 2025 in Australia.

ISBN : 978-1-7637778-8-0

All correspondence to: info@digitalcraftpublishing.com

www.digialcraftpublishing.com

DCP
DIGITAL CRAFT PUBLISHING

Hasan Baran Kaptan is an Australian-based manufacturing professional with more than a decade of experience spanning engineering, operations and digital-transformation roles. Trained as a mechanical engineer and later specialising in advanced manufacturing technology, he has worked for manufacturers in sectors ranging from automotive to consumer goods and advanced composites. Over the years he has held positions in production engineering, tooling and factory management before moving into digital-twin and Industry 4.0–focused roles.

Today Baran leads automation and digitalisation initiatives in a research-and-industry collaboration setting, helping organisations translate cutting-edge Industry 4.0 concepts into practical solutions. His work combines data-driven process improvement, lean manufacturing principles and the integration of sensors and software to create real-time digital representations of manufacturing processes. These digital twins enable better decision-making, predictive maintenance and more efficient operations.

With additional studies in the Industrial Internet of Things and lean practices, Baran brings a well-rounded perspective to manufacturing modernisation. He has a passion for demystifying complex technologies and guiding teams through change, making him a trusted adviser to engineers and leaders alike. In this book he distils his practical knowledge to help readers master the digital transformation of manufacturing.

The manufacturing industry stands on the brink of a profound transformation, driven by rapid advancements in digital technologies and automation. Today, concepts such as Industry 4.0, digital twins, artificial intelligence, and cyber-physical systems are no longer abstract ideas of the future but critical tools that define competitive advantage and operational excellence.

This book is designed to serve as a comprehensive guide to navigating the dynamic landscape of digital manufacturing. It demystifies key technologies, outlines actionable strategies for implementation, and highlights real-world examples of success. By bridging theoretical concepts and practical applications, the book provides readers with the knowledge and insights needed to successfully transition their manufacturing processes into highly connected, intelligent, and agile operations.

Whether you are a manufacturing professional seeking to stay ahead of technological trends, an industry leader driving digital transformation initiatives, or a student aspiring to shape the future of manufacturing, this book will equip you with the foundational understanding and practical guidance required to harness the full potential of digitalisation.

Welcome to the future of manufacturing.

Table of Contents

Introduction

Manufacturing is undergoing one of the most significant transformations in its history. Driven by rapid advancements in digital technology, this evolution, often described as Digital Manufacturing or Industry 4.0, fundamentally changes how factories operate, products are designed, and supply chains are managed. The shift from traditional, manual processes to interconnected, intelligent systems has profound implications for businesses, economies, and society.

This book aims to serve as a comprehensive guide to understanding and navigating this transformation. It explores key enabling technologies including Cyber-Physical Systems (CPS), Cloud and Edge Computing, Digital Twins, Manufacturing Execution Systems (MES), Product Lifecycle Management (PLM), Robotics and Automation, Data Analytics, Machine Learning (ML), and Artificial Intelligence (AI). At its core, the book provides readers with both a strategic overview and practical insights to embrace and effectively implement digital manufacturing practices.

Why now? Simply put, companies that fail to adapt to this digital transition risk falling behind. Recent research highlights that manufacturers who embrace digital transformation achieve significant competitive advantages, including up to 30% higher productivity, substantial reductions in downtime, and enhanced agility in responding to market demands. Given the accelerated pace of innovation and global competition, understanding and leveraging digital manufacturing is no longer optional—it's essential for sustainable growth and long-term success.

Importance in the Context of Industry 4.0

Industry 4.0 represents the fourth major industrial revolution, characterised by digital integration, data-driven decision-making, and extensive automation. Unlike previous industrial revolutions, which primarily introduced mechanical and electrical innovations, Industry 4.0 merges the physical and digital worlds. It integrates advanced information technology, operational technologies, and sophisticated analytics to create smart, connected manufacturing ecosystems.

Within this new landscape, digital manufacturing technologies drive unprecedented operational visibility, productivity, efficiency, and flexibility. Industry 4.0 empowers businesses to produce highly customised products at near mass-production costs, achieve unparalleled operational efficiencies through predictive maintenance and real-time optimisation, and rapidly respond to supply-chain disruptions through enhanced data insights.

Furthermore, embracing Industry 4.0 isn't merely about adopting new technologies—it also demands significant cultural and organisational shifts. It emphasises collaboration, continuous learning, and agility across all organisational levels. Manufacturers who strategically approach this revolution will be better equipped to tackle global challenges, meet evolving customer expectations, and secure their competitive positions for years.

Who Should Read This Book and How to Use It

This book is written for industry professionals, decision-makers, manufacturing engineers, technology strategists, operations managers, and executives who seek a clear, actionable understanding of digital manufacturing. It also provides value to consultants, researchers, educators, and graduate-level students interested in the practical implications of Industry 4.0.

To get the most from this book, readers should approach it both strategically and practically. Each chapter is structured to explain key concepts, highlight strategic implications, and offer valuable insights supported by current data and examples. While readers can navigate directly to chapters of particular interest, the book is designed as a coherent guide from foundational knowledge to advanced, strategic insights.

Begin by developing a comprehensive understanding of digital manufacturing's core concepts in the early chapters. Next, explore more profound insights and real-world applications in subsequent chapters, from CPS to AI. Finally, use the concluding sections to consolidate your understanding, assess future trends, and strategically plan your organisation's digital journey.

By engaging with the book's insights and examples, readers will be well-equipped to lead their organisations confidently into the future, turning digital transformation into tangible business value and competitive advantage.

Finally, I would like to personally thank you for taking the time to explore this book. Your willingness to invest your valuable time to enhance your knowledge and understanding of digital manufacturing is greatly appreciated. I sincerely hope this book will provide meaningful insights and practical guidance as you navigate your digital transformation journey. Your feedback is invaluable, and I warmly encourage you to share your thoughts and reflections after finishing the book. Your perspectives will help continually refine and enhance future editions, making this resource even more valuable to professionals like you.

Regards

Hasan Baran Kaptan

Digital Manufacturing

Digital manufacturing – often synonymous with "Industry 4.0" or smart manufacturing – is the integration of advanced digital technologies into every stage of production[1]. At its core are smart, connected systems that link machines, sensors, and software to monitor and optimise operations in real time[1, 2]. In practical terms, digital manufacturing uses tools such as the Industrial Internet of Things (IIoT), cloud computing, analytics, and automation to turn data into smarter factories. These systems can sense physical changes, analyse data on the fly, and even make decisions (or offer recommendations) instantaneously. For example, sensors throughout a plant can feed real-time performance data into analytics engines, enabling predictive maintenance and adaptive control without human intervention[1, 2]. By uniting digital design, production, and supply networks, digital manufacturing creates an end-to-end virtuous loop of continuous improvement.

Historically, manufacturing has evolved through successive "industrial revolutions" – from steam-powered mechanisation (Industry 1.0) to electrified mass production (2.0), to computer and robot-controlled automation (3.0). Today, we are living in the Fourth Industrial Revolution, a new era that blurs physical and digital worlds[3]. As World Economic Forum analysts note, "extraordinary technology advances" merge the digital, physical, and biological spheres, changing factories' operations [3]. The term Industry 4.0 itself was coined in Germany (circa 2011) to describe smart, connected production systems that can sense, predict, and interact with the environment[1]. Unlike past revolutions focused mainly on mechanical or electrical improvements, this wave is defined by pervasive connectivity (e.g. IoT networks) and advanced analytics (including AI/ML) throughout the value chain. In other words, factories are evolving from isolated, manual processes into cyber-physical enterprises.

Key Components and Technologies.

Digital manufacturing relies on an ecosystem of new technologies and cyber-physical components.

Industrial Internet of Things (IIoT) & Cyber-Physical Systems: Smart sensors and embedded controllers on machines and tools collect data and send commands over the network. These IIoT devices (actuators, robotics, wearables, etc.) link the physical plant to the digital world[4]. Together with traditional automation (PLC, CNC), they form cyber-physical systems (CPS) that monitor and adjust processes in real time.

AI and Data Analytics: Machine learning and AI algorithms analyse the vast data from the shop floor. They detect patterns, predict machine failures (predictive maintenance), and optimise operations on the fly. Early adopters of AI in manufacturing have seen measurable benefits – for example, one industry report notes that smart AI-driven analytics enabled companies to achieve roughly 14% cost savings by identifying inefficiencies[5]. By continuously learning from data, AI makes factories more proactive and adaptive.

Robotics and Automation: Robots (including collaborative robots or "cobots"), autonomous guided vehicles (AGVs), and automated machinery carry out repetitive or complex tasks. Robotics boosts speed, precision and consistency. Globally, the use of industrial robots has accelerated – factory robot density (robots per 10,000 manufacturing workers) hit a record 162 in 2023, more than doubling the level of 2016[6]. This reflects how automation is rapidly spreading across sectors.

Digital Twins and Simulation: Virtual replicas of products, machines or whole factories – known as digital twins – allow engineers to simulate and test designs before building them. Digital twins mirror real-world performance and enable "what-if" experiments without disrupting production. Analysts project that scaling up digital-twin technology could unlock enormous value (for example, an Accenture/WEF study estimates $1.3 trillion in global value and 7.5 gigatonnes of CO_2 savings by 2030 from top digital-twin use-cases)[7]. Thus, simulation tools are a key enabler of innovation and sustainability in manufacturing.

Enterprise Software (PLM, MES, ERP, etc.): On the software side, platforms like Product Lifecycle Management (PLM), Manufacturing Execution Systems (MES) and Enterprise Resource Planning (ERP) manage product data and production flows. PLM systems capture design and engineering data, MES coordinates shop-floor activities, and ERP handles supply chains and finance. Together, these systems provide the digital backbone. For example, standard textbooks note that MES, ERP and PLM serve respectively to coordinate production, manage resources, and oversee product development[4]. By integrating this enterprise software, manufacturers ensure that design updates, inventory decisions and production orders stay in sync.

Cloud and Edge Computing: Cloud platforms provide large-scale data storage and advanced processing power for manufacturing analytics, while edge computing brings real-time processing close to the machines for low-latency control. In practice, data from an IIoT sensor might be analysed at the edge for immediate feedback (e.g. shutting down an overheating motor) while longer-term trends are processed in the cloud. This hybrid computing approach underpins scalability and agility in a digital factory[4].

Human-Machine Interfaces (AR/VR, HMIs): The workforce interacts with the system via modern interfaces: digital dashboards, augmented/virtual reality (AR/VR) work instructions, mobile apps, etc. These interfaces help operators visualise data, receive alerts, and guide maintenance or assembly tasks. For example, augmented reality tablets can overlay schematics onto equipment for maintenance crews. Incorporating human-machine collaboration tools is essential to leverage human expertise and digital precision.

Collectively, these technologies form the digital manufacturing ecosystem. They are linked by the digital thread: product and process data flows across design, simulation, production planning, execution and inspection[8]. In essence, digital manufacturing transforms a factory into a cyber-physical system where software and machinery co-evolve.

Strategic Importance Today

Digital manufacturing is now a strategic imperative. Companies that digitise well are able to transform their cost structure, agility, and market position. Industry surveys show that 92% of manufacturers believe smart manufacturing will be their main competitive advantage driver over the next few years[9]. Leading adopters report dramatic gains: McKinsey finds that advanced factories can "increase production capacity and reduce material losses" while also improving service and speed[10]. For example, front-running companies often capture reductions in machine downtime (by 30–50%) and boost throughput by 10–30%[10]. These gains, scaled across a factory network, fundamentally transform a firm's competitive position[10]. In many industries, digital advantages translate directly into market share: reduced costs, faster time-to-market, and better customer responsiveness often come hand-in-hand with digital progress.

Governments and industry alliances worldwide recognise this. National programs from Germany's "Industrie 4.0" to China's "Made in China 2025" underscore that digital manufacturing is core to economic development. Even at the policy level, experts note that manufacturing "automates large parts of production" and tightens real-time ties to customers and markets[2]. In short, digitalisation is no longer optional – it underpins efficiency and innovation in modern manufacturing. As OECD analysts observe, digital technologies are "paving the road to Industry 4.0," enabling higher quality at lower cost while connecting complex supply chains[2]. By generating data-driven visibility into every process, companies can spot inefficiencies and respond swiftly – a critical edge in today's fast-changing markets.

Adoption Trends and Market Outlook

Adoption of digital manufacturing technologies is accelerating globally. Recent industry surveys and market analyses paint a picture of rapid growth:

Investment and Priority: Most manufacturers are pouring resources into digital initiatives. In a 2024 Deloitte survey, 78% of respondents reported dedicating over 20% of their improvement budgets to smart manufacturing, and 88% expect to increase or sustain this investment next year[9]. Similarly, nearly half of manufacturers plan to prioritise investments in factory automation hardware, and about a third will focus on sensors and AI tools[9]. This reflects a widespread conviction that digital technology is crucial for resilience and growth.

Technology Adoption: Adoption of enabling technologies is already significant and growing. For instance, 57% of manufacturers now use cloud computing, and 57% deploy data analytics in production[9]. Nearly half (46%) have implemented Industrial IoT (IIoT) networks to connect machines, and 42% are leveraging 5G or similar networks for high-speed connectivity[9]. Usage of AI/ML is more nascent but emerging: about 29% of firms report deploying AI at scale in factories, and many more are piloting AI or generative AI projects[9].

Performance Impact: These investments are yielding tangible returns. In practice, companies report 10–20% lifts in production output and similar gains in labour productivity, with associated capacity increases[9]. (See below for a summary of common performance improvements.) Encouraged by these results, firms are embedding digital tech into more processes. For example, almost half (49%) of manufacturers say they are building smart processes end-to-end today, and another 15% aim for complete digitisation by 2026[11].

Market Size and Growth: Market analyses forecast double-digit growth in digital manufacturing. One industry report valued the global digital manufacturing market at ~$388 billion in 2023, projecting it to more than double (~$911 billion) by 2030 (roughly a 16% CAGR)[12]. While estimates vary, all point to rapid factory expansion of software, services, sensors, and automation. This growth is driven by advanced economies and emerging regions (e.g., India, Southeast Asia), recognising digital manufacturing as a path to competitiveness.

These trends indicate that digital manufacturing is moving from pilot projects to mainstream strategy. As one expert summary notes, smart factories have finally reached a "tide-shifting" moment[9], with broad executive buy-in and escalating budgets.

Impacts on Operations and Processes

Digital technologies can revolutionise factory operations. By enabling data-driven control and automation, companies can achieve substantial performance uplifts. For example, leading manufacturers routinely report metrics like:

- Machine downtime: −30–50%[10]

- Labour productivity: +15–30%[10]

- Throughput (output rate): +10–30%[10]

- Forecast accuracy: +85% (i.e. much tighter demand and supply planning)[10]

These improvements stem from interconnected systems that continuously monitor equipment and processes. Sensors detect drift or faults in real time, triggering predictive maintenance before failures occur; analytics optimise production schedules and quality control; and digital workflows eliminate paper delays. As a result, factories can run closer to capacity with fewer defects and faster changeovers. One McKinsey study cites examples of 30–50% reductions in downtime and significant gains in throughput after deploying advanced analytics and machine learning in plants[10].

Digitisation also improves flexibility. Retooling a production line or shifting product mix could take days or weeks in a traditional setup. In a digital factory, simulations and digital twins allow rapid "virtual testing" of changes before implementation. Order changes can be absorbed quickly by connected assembly lines and configurable machinery. This agility leads to shorter lead times and the ability to handle more product variety. Indeed, companies have reported faster first-time correct launches and greater ability to meet custom orders thanks to digital control of processes[10].

Operational efficiencies translate directly into cost savings and capacity gains. For example, companies often achieve higher overall equipment effectiveness (OEE) with digital monitoring, a key metric that combines availability, performance, and quality. In one documented case, a plant implemented a digital transformation targeting a 10-point OEE increase; through real-time data dashboards and autonomous constraint resolution, the plant exceeded its goals and cut per-unit costs by over 30%[10]. While not every factory will see those exact numbers, the broad pattern is clear: digitally enabled operations run smoother, faster, and leaner.

New Business Models and Competitive Advantage

Beyond factory floor gains, digital manufacturing is reshaping business models. Connected products and analytics create new value propositions: manufacturers can shift from selling products alone to selling outcomes or services. A classic example is that a heavy equipment maker might use IoT sensors to guarantee uptime and charge customers based on usage or availability ("equipment-as-a-service"). Data-driven insights also enable predictive maintenance contracts, spare parts optimisation, and real-time support services – all of which open new revenue streams.

Industry analyses highlight this shift. As PwC notes, leading companies are launching "disruptive digital solutions such as complete, data-driven services and integrated platform solutions"[13]. These digital business models focus on generating additional revenue and deepening customer engagement. For instance, a manufacturer might bundle software analytics with its machines or offer a cloud platform where customers can monitor performance globally. Such platform-based offerings often reach customers in novel ways and create stickier relationships.

Digital manufacturing also enables mass customisation. With flexible, software-controlled production, factories can economically produce small batches or one-off products tailored to individual needs. In consumer industries, this means on-demand personalised products. In industrial markets, it means modules or variants can be assembled according to customer specifications with minimal reconfiguration. In effect, digital tools collapse the trade-off between customisation and cost: as advanced analytics coordinate production, companies can offer greater variety without the traditional overhead.

All of these shifts bolster competitiveness. Firms that harness digital models can outpace rivals both in cost structure and customer value. Survey data support this: 92% of manufacturers say smart manufacturing will be the primary driver of competitiveness in their industry over the next three years[9]. This means companies at the technological frontier often pull ahead in market share. According to McKinsey, scaled digitisation "can fundamentally transform a company's competitive position"[10]. Conversely, laggards risk being undercut by more efficient, agile entrants or being squeezed by evolving industry standards.

Sustainability and Environmental Impact

Digital manufacturing also brings essential sustainability benefits. Optimising resource use, reducing waste, and improving efficiency align well with environmental goals. Digitally enabled processes tend to have a "win–win" effect: cost and carbon savings go hand-in-hand. For example, better process control reduces scrap and defects (saving material) and lowers energy use (since machines run only as needed). Sensors and analytics can identify energy-hungry operations and reduce consumption automatically.

World Economic Forum case studies show significant results: one plastics manufacturer installed energy sensors and analytics and cut power consumption in a plant by roughly 40%, saving over $200,000 per year[1]. Across industries, factories implementing smart production report lower emissions and waste. As McKinsey observes, digital transformations have "win–win advantages" of reduced environmental impact through "lower emissions and reduced waste"[10]. In the same vein, digital twins – virtual models of real factories – help optimise layouts and flows for minimal resource use, as manufacturers test "green" scenarios without physical trials[14]. One WEF analysis notes that continuous optimisation via digital twins inevitably leads to reduced water usage, energy use, and material waste, shrinking the factory's ecological footprint[14].

These sustainability impacts are increasingly valuable. Regulatory pressures and consumer demand favour greener practices, and digital tools make compliance easier (e.g. by providing traceability and reporting). In effect, digital manufacturing enables a "circular" approach: data from each product and process feeds back into design and production to use less material and recycle more. Over time, this supports broader goals like carbon reduction and resource efficiency.

Workforce Transformation and Skills

The rise of digital manufacturing is transforming the workforce. New skills are required to design, operate, and improve smart factories. While automation will take over many routine tasks, it also creates new roles for human workers in analytics, programming, system integration, and advanced equipment maintenance. In practice, many manufacturers face a pressing talent gap: surveys find roughly half of companies have difficulty filling operations and planning roles, and 35% cite adapting workers to the "Factory of the Future" as a top concern[9]. Companies report particular shortages in roles like IoT engineers, data analysts, and operators trained on advanced machines.

To address this, firms are investing heavily in training and culture change. Upskilling and reskilling are common—companies run internal programs to teach workers how to use new digital tools, collaborate with data scientists, or manage connected assets. The right digital tools can also make jobs safer and more engaging: McKinsey notes that automation and analytics "lead to more empowered decision making; new opportunities for upskilling, reskilling, and cross-functional collaboration; better talent attraction and retention; and improved workplace safety"[10]. In other words, workers shift from repetitive tasks to monitoring, analysis, and optimisation roles.

The transition is a challenge. Adapting talent pipelines (from education and training institutions) and addressing change resistance are critical. Without action, companies risk inefficiencies and lost opportunities: for instance, the World Economic Forum warns that failing to upskill could result in missed growth and weaker global competitiveness[15]. However, with strategic workforce planning, digital manufacturing can create higher-skilled, higher-value manufacturing jobs. Experts argue that preparing a "digitally savvy" workforce is key to unlocking the full potential of the transformation.

The Future of Digital Manufacturing

Looking ahead, digital manufacturing points toward autonomous, hyper-connected factories and unprecedented flexibility. For example, nearly half of the manufacturers surveyed expect fully or partially autonomous ("lights-out") factories in the coming years[16]. Advances in AI and robotics will enable machines to self-organise production flows and maintain without human intervention, aside from oversight. Similarly, embedded analytics and digital twins will allow factories to self-optimise, continuously tuning processes as conditions change (new orders, supply shifts, energy prices, etc.).

At the same time, product personalisation will surge. Digitally integrated supply chains and additive manufacturing could deliver hyper-personalised goods on demand. Thanks to flexible platforms, consumers may get one-of-a-kind products produced on a mass-production scale. Over time, the boundary between design, production, and service will blur: products in use will feed data back to designers, while smart factories reconfigure instantly for new models or variants.

These trends form the backdrop for tomorrow's innovations. Digital manufacturing is moving toward ubiquitous connectivity (5G/6G), pervasive AI that orchestrates entire value chains, and even more collaborative human–robot "cobots" on the floor. By merging physical automation with virtual intelligence, future factories will be highly resilient and adaptive. In sum, digital manufacturing today lays the foundation for a future of autonomous factories, extreme customisation, and continuous learning. This chapter has sketched the strategic landscape; subsequent chapters will delve into specific technologies (AI/ML, robotics, digital twins, etc.) that enable this revolution.

Data in Manufacturing

Modern manufacturing is experiencing a digital transformation, often called Industry 4.0, where data has become a critical asset. Manufacturers now collect vast amounts of information from machines, sensors, processes, and IT systems throughout the production lifecycle[17]. Leveraging this data enables a shift from reactive decision-making to proactive and even predictive operations[17]. This section provides a detailed look at how data is defined, collected, analysed, and utilised in contemporary manufacturing, along with the benefits, challenges, and best practices for industry professionals aiming to maximise data's value.

Definitions and Categories of Manufacturing Data

Manufacturing data encompasses all information generated and used during the production of goods. It can be categorised into several key types:

Machine Data: Data produced by industrial equipment and control systems. This includes machine controller logs, PLC (Programmable Logic Controller) signals, CNC machine programs, and robot operational data. Machine data often captures states (on/off, operating mode), performance metrics (cycle times, throughput), and events (faults or alarms). It provides insight into how equipment is performing and is usually time-series based. For example, a CNC machine might generate data on spindle speed, feed rate, and error codes during operation.

Sensor Data: Data from sensors measuring physical or environmental conditions in the factory. Sensors can monitor temperature, pressure, vibration, humidity, power consumption, etc., on machines or in the facility. This overlaps with machine data but also includes additional IoT sensor deployments (e.g. a vibration sensor on a motor, or a climate sensor in a cleanroom). Sensor data is typically high-frequency and granular, used to monitor conditions and detect anomalies. It can indicate, for instance, if a bearing's vibration exceeds a threshold (predicting a failure) or if a furnace's temperature deviates from the setpoint.

Process Data: Data describing the manufacturing process execution and parameters. This includes production counts, process parameters (speeds, feeds, temperatures, voltages, recipes), cycle times, line throughput, and yield rates. Process data may be collected by MES (Manufacturing Execution Systems) or SCADA systems during production runs. It answers what was made, how, and under what conditions. For example, process data would include injection pressure, cooling time, and the number of parts moulded per hour in an injection moulding process.

Quality Data: Information related to product quality and compliance. This includes inspection results, measurements, defect rates, scrap counts, rework logs, and statistical process control (SPC) charts. Quality data might come from visual inspection systems, coordinate measuring machines (CMMs), or quality management systems. It is used to ensure products meet specifications and to trace issues. For instance, after a batch production, quality data could show that 98% of units passed testing while 2% failed due to a specific dimensional defect.

Supply Chain and Logistics Data: Data concerning the flow of materials and products into and out of the factory. This includes inventory levels, material traceability records, supplier information, order status, and shipping/delivery data. Such data often resides in ERP (Enterprise Resource Planning) or supply chain management software. It helps coordinate manufacturing with procurement and distribution, for example, by tracking that a critical raw material lot from a supplier was used in a specific production batch.

Other Relevant Data: Manufacturing also involves data from maintenance (e.g. maintenance work orders, machine downtime reasons), product design (design specifications, CAD files, BOM from

PLM systems), and even customer feedback (warranty claims, product performance in the field). All these data types can be considered part of the manufacturing data ecosystem. In essence, "big data" in manufacturing can refer to information collected from virtually any process or activity — machine sensor readings, quality assurance info, supplier data, maintenance logs, financial metrics, and more[18]. Each category provides a different lens on operations, and together they form a comprehensive picture of manufacturing performance.

Data Sources Across the Manufacturing Lifecycle

Data in manufacturing is generated across the entire product lifecycle, from design and planning to production and delivery. Key data sources include a variety of enterprise and shop-floor systems and devices:

Product Lifecycle Management (PLM) Systems: PLM software manages product design data, engineering specifications, and the product's development history. It is a data source like CAD models, part drawings, bills of materials (BOM), and engineering change orders. This design data informs manufacturing processes (e.g. dimensions and tolerances that machines must achieve) and can be linked to production data for traceability. PLM systems ensure that the latest design revisions are communicated to manufacturing and that any issues discovered in production can be fed back to engineering.

Enterprise Resource Planning (ERP) Systems: ERP is the central business system that handles production planning, materials management, supply chain, and finance. ERP provides data such as production schedules, work orders, inventory levels, procurement orders, and shipment logs. For example, an ERP system will house a production plan that tells the factory what products to make in what quantities by specific dates, and it will record the consumption of raw materials and the completion of finished goods. ERP data is critical for aligning manufacturing output with customer demand and tracking costs and deliveries.

Manufacturing Execution Systems (MES): MES operate at the plant floor level to execute and monitor production in real time. They track work-in-progress (WIP), dispatch work orders to machines or operators, record each step of production, and capture detailed process data and production events. MES are a rich source of process data and production records — for instance, they can record that on Line 3, Order #1234, 500 units were produced between 8 AM and 2 PM with Lot ABC of raw material, including any downtime or quality checks that occurred. MES is often linked to ERP (receiving orders and reporting completions) and control systems (gathering machine signals or sending setpoints).

SCADA and Industrial Control Systems: SCADA (Supervisory Control and Data Acquisition) systems and related control systems (DCS in process industries, or simply HMI/PLC setups in discrete manufacturing) are sources of real-time machine and sensor data. SCADA software provides a supervisory layer to collect data from PLCs, sensors, and actuators on the factory floor, and may store this data in a historian (described later) for trending. SCADA displays and logs events like sensor readings, alarms, and equipment status. For example, a SCADA system in a chemical plant might gather temperature, pressure, and flow data from various points in a reactor and allow operators to supervise the process. SCADA is a key source for granular time-series data and is often integrated with MES to provide a bridge between the shop floor control and higher-level execution systems.

Industrial IoT Devices and Sensors: Additional IoT devices supplement traditional control systems in modern smart factories. These include wireless sensors, smart meters, and edge IoT gateways that collect data that might not have been captured before. IoT devices can be attached to legacy equipment to retrofit them with new sensors (for vibration, energy use, etc.), or monitor environmental conditions (air quality, temperature in different facility areas) and asset location (via RFID, for instance). They often transmit data via IoT protocols (like MQTT) to cloud platforms or edge servers. The proliferation of IoT means manufacturers can measure a "wide range of aspects of the manufacturing environment, from individual machine performance to overall production outputs" with relative ease[19]. These devices dramatically increase the volume and variety of data available.

Robotic and Automation Systems: Industrial robots, automated guided vehicles (AGVs), and other automated machinery also produce data. Modern robots are equipped with controllers that log information such as cycle counts, positions, payload weights, error codes, and even internal sensor data (torque, vision system outputs). Integration of robotics data can be through direct network connections or via the MES/SCADA if they interface. For example, a pick-and-place robot might provide data on how many picks it has done, any missed picks, or mechanical stress readings that could predict maintenance needs. Similarly, automation systems like conveyor PLCs or packaging machines provide data on throughput and jams. All of this falls under machine data but is worth noting separately since automation systems are ubiquitous data sources in manufacturing.

Quality and Laboratory Systems: Although not explicitly listed in the prompt's examples, many factories have dedicated quality management systems (QMS) or laboratory information management systems (LIMS) that handle inspection data, test results, and certifications. For completeness, these systems source data such as dimensional inspection results, metallurgical test outcomes, or software test logs (for electronics). They contribute to the quality data category and often need to link with production data for full traceability (e.g. tying a failed test back to the exact production conditions or material lot).

Maintenance Management Systems: Another ancillary source is computerised maintenance management systems (CMMS) that log machine maintenance activities, schedules, and repairs. Maintenance logs (when a machine was serviced, what parts were replaced), combined with machine sensor data, can feed into reliability analysis and predictive maintenance models.

Across the lifecycle, the goal is to integrate these data sources so that a "digital thread" connects product design, production, quality, and supply chain. For instance, data might flow from PLM (design specs) to MES (to enforce the correct process for that design), from MES/SCADA to quality systems (to analyse if process parameters affected quality outcomes), and from MES back up to ERP (to update inventory and order status). By capturing data at each stage, manufacturers build a comprehensive record of the product lifecycle – from initial design requirements to the moment a product ships (and even beyond, if field data from products in use is collected). Modern plants often rely on an integrated architecture (following standards like ISA-95, discussed later) to ensure these diverse systems can exchange information seamlessly.

Methods of Data Collection, Transmission, and Storage

Collecting and managing manufacturing data requires robust methods for gathering it from the source, transmitting it where it needs to go, and storing it for analysis and historical reference. Key aspects include edge vs. cloud processing, use of data lakes and warehouses, and specialised industrial data repositories. Below are essential methods and technologies for data collection, transmission, and storage:

Edge Computing: Edge computing refers to processing data near the source (at the "edge" of the network, such as on the factory floor devices or local servers) rather than sending everything to a central cloud or data centre. In manufacturing, edge computing is often implemented via industrial PCs, IoT gateways, or even on-controller computing resources. The advantage is reduced latency and bandwidth usage, which are critical for real-time control or analytics. For example, an edge gateway might sit on a machine network, directly ingest high-frequency sensor data (vibrations, currents, etc.), perform first-line analytics or filtering (such as detecting an anomaly or aggregating data), and then send only summarised or relevant data to the cloud. Edge computing can also enable continued operation if internet connectivity is lost (the local edge keeps working). This approach is commonly used for applications like real-time machine monitoring or vision systems, where the volume of raw data (e.g. video feed) is enormous and immediate response is needed; processing that is local and only sending results (like "part OK" or "part NOT OK") to central systems is more efficient.

Cloud Storage and Cloud Computing: Cloud platforms provide scalable storage and computing for manufacturing data. Many manufacturers are aggregating plant data into cloud data lakes or time-series databases to perform advanced analytics across sites. Cloud storage is virtually unlimited and can handle the "massive volume and velocity" of data that modern facilities generate[20]. Data from IoT sensors and MES might be published to cloud services (using MQTT or HTTP APIs) where it's stored for analysis or machine learning model training. Cloud computing also facilitates combining manufacturing data with other enterprise data (like supply chain or customer data) to gain broader insights. For instance, a company could stream machine sensor data to an IoT cloud (like Azure IoT or AWS IoT), apply analytics or AI models there to predict failures, and then visualise results in a cloud dashboard accessible from anywhere. While cloud computing offers flexibility and powerful analytics tools, manufacturers must consider network reliability, latency (for any control decisions, latency might be too high to do from the cloud), and security of data in transit and at rest.

Data Lakes and Warehouses: A data lake is a centralised repository that stores raw data in its native format (structured, semi-structured, unstructured) at scale. In manufacturing, data lakes are used to consolidate information from various sources – MES logs, sensor streams, ERP tables, etc. – into one large pool where data scientists and analysts can then retrieve and analyse it. Because manufacturing data can be very diverse (numbers from sensors, text from operator logs, even images from vision systems), a data lake is helpful for storing everything without requiring an upfront schema. Companies might implement a data lake using big data technologies (Hadoop/Spark) or cloud storage (e.g. AWS S3 with Athena/Redshift Spectrum, Azure Data Lake, etc.). On top of data lakes, data warehouses or analytics databases may be built to allow structured querying and business intelligence; for example, summarising daily production rates and yields per line for reporting. The data lake concept aligns with the idea of breaking down silos: all data goes into one hub accessible for many purposes. However, without proper governance, a data lake can become a "data swamp" – so metadata management and data quality processes are essential. Many manufacturers now create Industrial Data Lakes that combine time-series process data with transactional data to enable advanced analytics like digital twin simulations or multi-factor quality analysis.

Industrial Data Historians: Long before "big data" and lakes became popular, the manufacturing sector used (and still uses) data historian systems. A historian is a specialised time-series database optimised for high-speed capture and retrieval of time-stamped data from sensors and control systems[21]. Examples include OSIsoft PI System, Aveva Wonderware Historian, GE Proficy Historian, etc. These systems interface directly with PLCs, SCADA, and DCS to log process values (pressures, temperatures, motor currents, valve positions, etc.) typically at frequent intervals (e.g. once per second or faster). Historians efficiently compress and store this data, and allow querying or trending it easily (answering questions like "what was the temperature profile of Batch #102 on Line B last Tuesday?"). They are vital for process optimisation and troubleshooting, as engineers rely on historian trends to see how conditions have evolved. Many manufacturers consider the historian the backbone of plant data; however, as digital transformation progresses, historian data is now being integrated with other data via enterprise analytics platforms. A company in early digital maturity might not even have consolidated their time-series data into a historian yet, which is a foundational step before more advanced analytics. In modern architectures, historical data can be forwarded to data lakes or the cloud in addition to being available on-prem. The historian remains crucial for real-time and historical analysis on the shop floor, due to its reliability and speed for time-series queries.

Data Transmission Technologies: Underlying collection and storage is the matter of how data moves from point A to B. Manufacturing environments use a mix of industrial Ethernet networks, fieldbus protocols (PROFINET, EtherNet/IP, Modbus TCP, etc.), and wireless networks to transmit data. Increasingly, publish/subscribe messaging protocols like MQTT (Message Queuing Telemetry Transport) are employed to stream data from machines to central systems in real-time. MQTT's lightweight design minimises bandwidth and is well-suited for low-power or unreliable networks[22]. Traditional polling via OPC DA/UA or proprietary protocols is being complemented or replaced by these modern methods. Additionally, the rise of 5G private networks in factories promises high-bandwidth, low-latency wireless data transmission that could connect a multitude of sensors and devices without cables. Regardless of the medium, manufacturing data transmission must be reliable and often deterministic for control data. Thus, architectures usually segregate control networks (real-time critical, staying on-prem) from higher-level data networks (which can buffer and resend if needed). Technologies like OPC UA over TSN (Time-Sensitive Networking) are emerging to support time-critical data exchange in a standardised way.

In summary, modern manufacturers deploy a hybrid data architecture: edge computing for immediate, on-site needs; cloud and data lakes for big-picture analytics and cross-facility intelligence; historians for core process data storage; and robust networking protocols to tie it all together. This ensures that raw data from the shop floor is collected and contextualised, transmitted securely (whether to on-prem servers or cloud), and stored in a way that can support both instantaneous decisions and long-term strategic analysis.

The Role of Data Analytics in Manufacturing

Collecting data is only the first step – the real value comes from analysing it to derive actionable insights. Manufacturing analytics can be categorised into four escalating levels of sophistication: descriptive, diagnostic, predictive, and prescriptive analytics[23]. Each type serves a different purpose:

Analytics Type	Purpose & Questions Answered	Application in Manufacturing
Descriptive	What happened? Summarise and visualise historical or real-time data to understand events and trends.	e.g. Real-time dashboards and reports showing production counts, machine uptime/downtime, and quality metrics. Descriptive analytics in a factory might be an OEE (Overall Equipment Effectiveness) report that describes yesterday's performance on each line.
Diagnostic	Why did it happen? Dig deeper into data (often by correlation or drill-down) to find root causes of issues or deviations.	e.g. Root cause analysis of a quality issue by examining process parameters and sensor data leading up to a defect. If a batch had low quality, diagnostic analytics might reveal that a temperature sensor drifted out of range at a critical time, explaining the issue.
Predictive	What will happen? Use statistical models and machine learning on historical data to forecast future events or probabilities.	e.g. Predictive maintenance: analysing vibration and temperature trends to predict when a machine is likely to fail. Likewise, predicting future product demand or detecting a trend toward increasing defect rates. For instance, a predictive model might forecast that a particular motor has an 80% chance of failing in the next 10

		days, allowing planners to schedule a replacement.
Prescriptive	What should we do about it? Recommend actions or optimisations based on predictive insights and objectives, often using AI optimisation or simulation.	e.g. Automatic adjustment of process settings to optimise performance (self-tuning). A prescriptive analytics system could suggest an optimal production schedule or optimal machine speed settings to minimise energy use while meeting output targets. In some cases, prescriptive solutions directly implement changes (closed-loop control), such as an AI system that fine-tunes a chemical recipe in real-time to maximise yield.

Table: Type of analytics and their uses in manufacturing

In manufacturing, organisations typically start with descriptive analytics (basic monitoring and reporting) and progress towards more advanced types as their data maturity grows. Descriptive and diagnostic analytics are often part of daily operational excellence – for example, monitoring dashboards that alert if KPIs stray from target, and investigating any anomalies. They rely on good visualisation and query tools, often built into MES or BI (Business Intelligence) platforms.

Predictive analytics and prescriptive analytics are considered "advanced analytics." They usually require more complex tools: machine learning algorithms, big data platforms, or digital twin simulations. These can unlock significant improvements – for instance, predictive models can anticipate equipment failures so that maintenance can be performed just-in-time, avoiding unplanned downtime[24]. Prescriptive analytics might run thousands of "what-if" scenarios (sometimes via a digital twin of the factory) to identify the best course of action under certain

31

conditions (e.g. how to re-route production if a machine goes down). Organisations that successfully implement predictive/prescriptive analytics often see a step-change in performance, moving from reactive firefighting to proactive optimisation.

To illustrate the power of advanced analytics, one report notes that data-driven strategies in manufacturing have led to 10–15% productivity improvements and 15–20% cost reductions on average[25]. These gains come from things like reduced downtime, better yields, and more efficient resource use – all enabled by analysing data and acting on the insights. However, achieving this requires not just tools but also that all relevant data is aggregated and accessible to the analytics systems[23]. Many manufacturers find that they need to first integrate their disparate data (from historians, MES, ERP, etc.) into a unified platform before they can fully leverage predictive or prescriptive analytics.

In summary, data analytics in manufacturing spans from reporting the basics to using AI for decision automation. Each level builds on the previous: descriptive analytics might show that a machine's output dropped, diagnostic finds it was due to a specific fault, predictive warns that such a fault is likely to recur next week, and prescriptive suggests ordering a replacement part and reassigning tasks to other machines in the meantime. Together, these analytics capabilities enable smarter, faster decision-making on the shop floor and in management.

Use Cases and Benefits of Manufacturing Data

When manufacturing data is effectively collected and analysed, it unlocks numerous high-impact use cases. Industry professionals are leveraging data to optimise processes, improve quality, reduce downtime, save energy, and ensure traceability. Here are some of the significant use cases and their benefits:

Process Optimisation and Efficiency Improvement

Data-driven process optimisation is about using insights from production data to improve throughput, reduce cycle time, and eliminate bottlenecks. By analysing machine utilisation and production flow data, manufacturers can identify inefficiencies in their operations. For example, monitoring data might reveal that one machine on an assembly line is frequently idle or is waiting on upstream parts, indicating a bottleneck. Using such insights, the production schedule can be adjusted or the process rebalanced. Studies show that analysing data about machine usage highlights which machines are bottlenecks, which are underutilised, and which are being overstrained[18]. With this knowledge, managers can take action such as redistributing workload, adding parallel equipment, or improving maintenance on an overburdened machine.

Another aspect of optimisation is adjusting process parameters to maximise output. In a data-rich environment, manufacturers perform experiments or use AI to find optimal settings. For instance, a plant might use historical data to determine that running a machine at 90% of its maximum speed yields better overall throughput (with less downtime) than running it at 100%, where it fails more often. Process mining and analysis can also reduce changeover times, optimise batch sizes, and minimise waste. The net benefit of process optimisation is higher productivity and often lower cost per unit. Data-driven decisions remove a lot of the guesswork – instead of relying on tribal knowledge, companies use evidence from data to fine-tune their operations continuously.

Quality Improvement and Defect Reduction

Quality control and assurance greatly benefit from manufacturing data. By capturing data at each production step and during final inspection, manufacturers can correlate conditions with quality outcomes. For example, suppose data shows that products made on a Monday morning shift have a higher defect rate. In that case, one can investigate further (maybe a calibration drift over the weekend, or a particular operator training issue). Diagnostic analytics on process vs. quality data often uncovers root causes of defects – perhaps a certain temperature range produces weaker welds, or a specific material lot from a supplier caused issues.

Advanced techniques like machine learning are also being applied for quality improvement. Vision systems and sensors generate large datasets used to train models that can detect subtle anomalies in products that humans might miss. Additionally, sentiment analysis on customer feedback and field performance data (a form of external data) can identify recurring quality issues[18], which then drive improvements in manufacturing processes to address those pain points.

Crucially, real-time quality data allows for immediate adjustments. If an in-line sensor detects a dimension drifting toward an out-of-spec value, the MES or control system can trigger an alert or even auto-adjust the process (if prescriptive control is in place). This prevents defects rather than just detecting them after the fact. The benefit is a higher first-pass yield and fewer scrapped or reworked parts. Data also helps in implementing Six Sigma and statistical process control methodologies by providing a continuous stream of measurements needed to analyse variance and keep processes centred.

Predictive Maintenance and Asset Reliability

One of the most celebrated use cases in Industry 4.0 is predictive maintenance. Traditionally, manufacturers either serviced equipment on fixed schedules or ran them to failure. Both approaches have downsides: too frequent maintenance wastes time and parts, while unexpected breakdowns cause costly downtime. Predictive maintenance uses data (like vibration, temperature, acoustic signals, lubricant analysis, machine error codes) to predict when equipment is likely to need maintenance before a failure occurs[24, 26]. By applying algorithms to historical failure data and real-time condition monitoring, manufacturers can identify patterns that precede failures.

The benefits of predictive maintenance are significant. It minimises unplanned downtime by addressing issues at convenient times. According to the U.S. Department of Energy, predictive maintenance can reduce maintenance costs by 25–30% compared to reactive strategies[24]. It also extends equipment life by avoiding catastrophic failures and overusing machinery. For example, an automotive supplier using sensor data might forecast that a robot arm's motor will fail in 10 days – they can replace that motor over the weekend rather than having it break mid-week, halting production. This data-driven approach improves equipment availability and line reliability, directly boosting OEE and throughput.

Beyond preventing breakdowns, the data used for maintenance can also improve safety (detecting hazards like abnormal vibrations that could lead to accidents)[24] and optimise spare parts inventory (knowing which parts will likely be needed so that they can be ordered just in time). Many companies start their digital journey with predictive maintenance as a clear ROI use case, then expand similar predictive methods to other areas like quality and supply chain.

Energy Efficiency and Sustainability

Energy costs are a significant factor in manufacturing, and there is growing pressure for sustainability. Data helps monitor and reduce energy consumption. By collecting data from power meters, equipment PLCs, and facility systems (HVAC, compressors, ovens, etc.), manufacturers gain visibility into where energy is used and wasted. Energy analytics can pinpoint, for instance, that a particular machine draws a lot of power during idle states, suggesting a need for better standby modes or shutdown policies.

With data, companies can perform load optimisation – scheduling high energy-use processes at off-peak hours, or staggering the startup of machines to avoid demand spikes. They can also detect anomalies like compressed air leaks or inefficient motors via energy signatures. Using predictive analytics on energy data can identify inefficiencies (e.g. a machine using more power than usual due to a developing fault) so that it can be fixed, thereby reducing energy waste(24). Moreover, data-driven process optimisation often concurrently improves energy efficiency; for example, eliminating a bottleneck could reduce the time other machines run empty, saving energy.

Some manufacturers integrate energy data with production data to calculate energy per unit produced, a key metric for sustainability. By tracking this, they set targets and use data to drive improvements (like adjusting oven temperature profiles or cooling flow rates). The benefits include lower utility costs and a smaller carbon footprint. Real-time energy monitoring can also help in demand response programs (adjusting consumption in response to grid conditions), which can bring incentives. In summary, data empowers manufacturers to produce more with less energy by shining a light on exactly how and where energy is consumed in the production process.

Traceability and Supply Chain Visibility

Traceability is the ability to track every component and material from origin through manufacturing and into the final product, and to trace finished goods out through distribution. Data is the backbone of effective traceability systems. In practice, traceability involves capturing identifiers (like batch numbers, serial numbers, and timestamps) at each step and linking them. For example, when a raw material lot is received, its barcode is scanned into the system; when that lot is used in production, the system logs which work order and machine it went into; later, the final product gets a serial number that is linked to that work order and hence to the raw material lot. All these events generate data records.

Modern MES and ERP systems facilitate this by recording genealogies (which sub-parts went into which assembly, etc.). Data traceability allows a manufacturer to know the origin, history, and destination of any product or component through unique identifiers like barcodes or RFID tags[27]. The benefits of this are manifold:

Quality control and recalls: If a defect or contamination is discovered, traceability data can isolate the scope. Manufacturers can pinpoint exactly which batches or serial numbers are affected instead of recalling everything. In the event of a recall, a robust traceability system lets the company quickly identify and pull only the affected products, minimising consumer risk and protecting the brand[27].

Regulatory compliance: Many industries (food, pharmaceuticals, aerospace, automotive) have strict regulations requiring proof of traceability. Digital traceability data demonstrates compliance by showing the chain-of-custody and manufacturing conditions for each item[27]. This avoids legal penalties and ensures market access.

Supplier accountability: If a problem is traced to a supplier's material, the data proves it, enabling corrective action with that supplier. Similarly, if a supplier needs to be informed of an issue, traceability provides the exact lot numbers involved.

Process improvement: Traceability data, when analysed, can reveal patterns (e.g. components from Supplier X consistently result in fewer failures than those from Supplier Y, or a specific production line yields better results for a particular product). This can drive decisions in sourcing and production routing.

Additionally, extending traceability into the supply chain via data sharing provides end-to-end visibility. For instance, IoT trackers on shipments can feed location and condition data (temperature, shock) back to a central system, so manufacturers know in real time where their goods are and if they remain within specified conditions. Overall, traceability data builds a foundation of trust and knowledge, both for internal process control and external stakeholder assurance, by ensuring every product's story can be told from raw material to customer.

Integration with Enterprise Systems and Digital Twins

To fully exploit manufacturing data, it must flow freely between various systems. Integration is about breaking down data silos and enabling a seamless exchange of information between the shop floor (OT – Operational Technology) and the top floor (IT – Information Technology). Key integrations include MES with ERP and PLM, as well as the incorporation of digital twin technology.

MES–ERP Integration: MES and ERP integration are often aligned with the ISA-95 architectural model (Level 3 MES connects to Level 4 ERP). The ERP system plans what and when to produce, while the MES executes and monitors how it's produced. Integrating the two means that as soon as ERP creates a production order, it's sent to MES for scheduling on specific machines. Conversely, as MES tracks production progress (quantities made, downtime events, quality results), it feeds that back to ERP in near real-time[28]. This closes the loop: inventory in ERP is updated as soon as units are made, and customer orders can be committed based on actual production status. A well-integrated MES-ERP setup eliminates manual data entry and delays, improving the accuracy of promise dates and inventory records. For example, if an MES records that a batch is completed and put in finished goods stock, the ERP can automatically see that those goods are available for shipment. Without integration, such updates might happen only at the end of each shift or through error-prone manual reports.

PLM and Engineering Integration: Integration between PLM (design), MES (execution), and ERP (planning) ensures that the latest product definitions and process plans are communicated to manufacturing, and that manufacturing can provide feedback to engineering. This can involve transferring BOMs and routings from PLM/ERP into MES, so that MES knows the required steps and components for a new product. Additionally, if a design change occurs (say a part is updated to a new revision), integration propagates this change to production orders and work instructions, preventing the use of obsolete parts. On the feedback side, MES can send quality data or engineering change requests back to PLM if, for instance, a design doesn't fit the production reality. This integration is part of the concept of a digital thread, where shared data connects all lifecycle stages. The result is faster time-to-market (since manufacturing is involved early with precise data) and fewer errors (since everyone references the same single source of truth for specifications).

SCADA/PLC – MES Integration: At the control level, machines and PLCs often communicate up to MES or IIoT platforms using protocols or middleware (like OPC UA servers or MQTT). This integration means that MES can automatically fetch production counts and machine states and even trigger actions. For instance, when a work order starts, MES might send a signal to a line to download the correct machine program or to set a recipe parameter; likewise, when a machine finishes a cycle, it might send a signal that increments the count in MES. Integration here reduces the need for operators to input data manually and provides real-time visibility. It also enables automation of processes, such as automatically halting a machine when a specific count is reached (because the order quantity is done) or when a quality check is due.

Digital Twin Integration: A digital twin is a virtual replica of a physical entity (a machine, production line, factory, or even the entire supply chain) that is kept in sync with real-world data. Digital twins in manufacturing rely on integration with all the aforementioned systems to stay up-to-date. For example, a factory digital twin will ingest data feeds from machines (sensors, PLC data), MES, ERP, and even operator inputs (HMI data) to mirror the state of the factory in real time. By integrating these data streams, the digital twin can simulate outcomes and support decision-making. McKinsey describes different scopes of twins: an asset twin might integrate with PLC and sensor data for one machine to enable predictive maintenance; a factory twin composes data from assets, MES, ERP, and more to allow holistic optimisation (like dynamic scheduling or layout changes)[29].

The benefit of integrating digital twins is that manufacturers can test "what-if" scenarios in the virtual world before implementing them physically. For instance, the twin could use real production data to simulate what would happen if a particular machine's speed is increased by 10% or if a new product variant is introduced on the line. This helps identify potential issues or optimal settings without disrupting actual operations. Digital twins essentially act as high-level prescriptive analytics tools that are only as good as the data fed into them, hence the emphasis on integration. As noted, complete factory digital twins are enabled by data feeds from virtually every system (assets on the shop floor, MES, ERP, human interfaces) working in concert.[29]

Breaking Down Silos: Underlying all integration efforts is the goal of eliminating data silos. In many traditional factories, the PLM, MES, ERP, and equipment control systems functioned in isolation or with minimal point-to-point interfaces. This made it hard to get a unified view. Modern integration approaches often use middleware or IIoT platforms that subscribe to all data and then redistribute it where needed (sometimes called a Unified Namespace in IIoT architecture). Interoperability is enhanced by using standard data models and protocols (like OPC UA for machines or B2MML/ISA-95 for MES-ERP). The ISA-95 standard specifically "provides a framework for integrating enterprise and control systems"[28], defining what information should flow between levels. Many vendors and open-source solutions implement these standards to simplify integration.

In practice, successful integration results in capabilities like end-to-end traceability (PLM → MES → ERP → CRM), real-time production monitoring in corporate dashboards, automated lot genealogy (tying materials from ERP to processing details in MES to quality results in QMS), and synchronisation between multiple plants. It also lays the data foundation required for advanced analytics and AI (since those often need data that resides in different systems). As a concrete example, consider an organisation that integrates a machine condition monitoring system with its maintenance management and ERP: a vibration alert on a machine could automatically create a maintenance work order in the CMMS and check the ERP for spare part availability, all without human intervention. Such integration is the hallmark of a truly smart factory.

Challenges in Managing Manufacturing Data

While the promise of data-driven manufacturing is great, there are several challenges that companies must address to successfully harness data. These challenges include technical, organisational, and security issues:

Data Silos and Integration Difficulties: Many manufacturers find that their data is isolated in different systems or departments. For instance, production line data might be stored in a historian, quality data in spreadsheets, and inventory data in an ERP, with little connection between them. When data is kept in separate systems and formats, it impedes comprehensive analysis and decision-making. Siloed data leads to situations where engineers cannot easily correlate quality issues with specific machine settings because the data is stored in different databases. The lack of integration means missed insights. Breaking down silos often requires significant effort in systems integration (using middleware, APIs, or IoT platforms) and standardising data formats. Until that happens, organisations struggle to get a "single source of truth." This challenge is common in the early stages of digital transformation, where different functions (production, maintenance, supply chain) have optimised their IT systems but not the cross-talk between them.

Interoperability Issues: Related to silos, interoperability is the ability of different machines and software to exchange and use information seamlessly. Factories often have heterogeneous equipment – different makes, models, vintages – each with its data protocols. Older machines might output data only in proprietary formats or not at all, making it hard to integrate them. The lack of standardised data models and protocols can hinder seamless integration, effectively causing data fragmentation across devices and software. Standards like OPC UA and MQTT (discussed in the next section) aim to improve this, but adoption can be inconsistent. Achieving interoperability may require retrofitting older equipment with IoT gateways or using translators that convert one protocol to another. Additionally, semantic interoperability (ensuring that data labels/units have consistent meaning across systems) is a challenge – another system might not automatically recognise one system's "Temperature" tag without mapping and context. Companies often need to create a unified data dictionary or model (e.g., via ISA-95 or digital twin models) to ensure that once the data is connected, it is logical and usable.

Cybersecurity Risks: As factories become more connected (machines to MES, MES to cloud, remote access enabled, etc.), the attack surface for cyber threats increases. Manufacturing has become a common target for cyberattacks ranging from ransomware to state-sponsored industrial espionage. Each new connection point between systems, whether a sensor feeding data wirelessly or a cloud integration, can be a potential entry for attackers[30]. Cyber incidents can result in stolen intellectual property, production sabotage, or safety incidents if critical systems are tampered with. Thus, ensuring data security is paramount. This is challenging because operational technology was historically isolated (air-gapped networks) and not designed with cybersecurity in mind. Now, manufacturers must implement robust network security (firewalls, segmentation of IT and OT networks), data encryption in transit, authentication and authorisation for devices and users, and regular security updates – all without disrupting operations. Security monitoring specific to OT (for example, detecting unusual PLC commands) is also needed. The challenge is technical and cultural: OT personnel may not be used to frequent patches or downtime for security, and IT personnel may not understand the legacy protocols in use. Bridging that gap is part of the cybersecurity challenge in manufacturing data strategy[20].

Data Quality and Accuracy: The usefulness of data is only as good as its quality. In a complex manufacturing environment, data quality issues can arise from sensor errors (faulty or uncalibrated sensors giving wrong readings), manual data entry mistakes, missing data (e.g., network outages causing gaps), or inconsistent data (different systems using different naming or units). If data inputs are inaccurate or inconsistent, the outputs of analytics – the insights – will also be flawed ("garbage in, garbage out")[31]. Ensuring data accuracy in real time is challenging[20]. It requires validation rules, redundancy (multiple sensors to cross-check a critical parameter), calibration routines, and sometimes data cleaning procedures. For example, an IoT system might need to filter out spikes from a noisy sensor signal or interpolate missing values for a dashboard to be meaningful. Metadata (data about the data) is essential too – knowing units, timestamps, and context. Poor data quality can lead to incorrect decisions, such as a predictive maintenance system scheduling unnecessary maintenance due to a sensor glitch indicating high vibration that isn't actually happening. Therefore, part of data management is instituting data governance and quality control on datasets, much like product quality control.

Real-Time Processing Constraints: Manufacturing often requires real-time or near-real-time data processing. Data may need to be acted on in milliseconds or seconds for critical control decisions. The sheer volume and velocity of data, especially in IoT-enabled plants, can overwhelm traditional IT systems[20]. A single advanced machine tool can generate thousands of data points per second. Aggregating data from hundreds of devices can result in extremely high throughput requirements. If the architecture isn't up to it (insufficient network bandwidth, slow databases), decision loops will have latency. For example, suppose a system is analysing sensor data to detect anomalies, but the analysis comes several minutes late due to a backlog. In that case, it might miss the window to prevent a machine failure. Real-time constraints also mean that the system has to prioritise what data to process immediately at the edge versus what can be batch processed later in the cloud. Achieving accurate real-time analytics might require specialised stream processing frameworks and carefully engineered networks. Additionally, some legacy control systems are not designed to share data in real time with external systems, so tapping into them can be delicate (so as not to disturb their primary control task). The challenge is architecting a solution that can continuously ingest and process high-speed data without crashing or lagging. Scalability is related here, as operations expand or more sensors are added, the infrastructure must scale accordingly or risk slowdowns.

Organisational and Skill Barriers: Beyond the technical ones above, human challenges are worth noting. Often, manufacturing teams may not have the skill sets initially to manage big data or advanced analytics – data science and IT-OT convergence skills are in short supply. There can be resistance to change; operators or engineers might distrust algorithms or fear job displacement. Getting buy-in for data initiatives and training staff to use new tools is also a significant challenge. Building a data-driven culture (addressed in Best Practices) is how this is overcome, but it takes time and leadership.

Each of these challenges needs to be addressed for a successful data strategy. Many companies start by focusing on one (e.g., solving data silos by implementing a unified data platform), but all are interrelated. For instance, integrating systems (to fix silos) might introduce security concerns if not done carefully; pushing for real-time data might strain quality if the system can't validate it quickly enough. Therefore, a holistic approach is required, often under the umbrella of a strong data governance program.

Standards and Protocols for Industrial Data

Several standards and communication protocols have been developed and widely adopted to tackle interoperability and ease data exchange in manufacturing. These standards provide common languages and frameworks so that different devices and software can work together. Key ones include:

OPC UA (Open Platform Communications Unified Architecture): OPC UA is a platform-independent, service-oriented architecture for industrial communication. It is often called the universal translator for factory devices. OPC UA is more than just a protocol – it is a framework for industrial interoperability based on common data models[32]. In practice, OPC UA allows machines from different vendors to expose their data (like temperature, speed, pressure readings, and statuses) in a standardised way. Clients (like SCADA systems, MES, or IoT applications) can connect to an OPC UA server on a machine to read or subscribe to that data and even securely send commands. One of OPC UA's strengths is that its built-in security, encryption, and authentication are core parts of the design[32]. Also, it is extendable: industries create OPC UA Companion Specifications that define standard object models for specific types of equipment (for example, a companion spec for injection moulding machines will define what data and methods such a machine should expose). By implementing OPC UA, manufacturers and vendors significantly ease the integration effort. Instead of custom drivers for each machine, a standard OPC UA client can talk to any OPC UA-compliant server. It effectively helps unify the "meaning and description of the data" across systems[32]. OPC UA is often used at the interface of Level 2/3 of ISA-95 (machine control layer to MES layer) and is a cornerstone of many Industry 4.0 architectures.

MQTT (Message Queuing Telemetry Transport): MQTT is a lightweight publish/subscribe messaging protocol commonly used in IoT. It was designed for low-bandwidth, high-latency or unreliable networks and devices with limited resources. MQTT provides reliable messaging with minimal network overhead and code footprint[22]. In an MQTT model, devices (sensors or equipment) publish messages on topics (essentially labelled channels) to an MQTT broker, and any system interested (subscriber) that has subscribed to those topics receives the messages. This decouples producers and consumers of data – they don't need to know about each other, only share the topic naming convention. MQTT is widely used in manufacturing to stream data from edge devices to the cloud or central systems. For example, each machine could periodically publish its data to a topic like factory1/line2/machineA/temperature, and a cloud analytics engine subscribed to factory1/+/+/temperature gets all temperature readings from factory1. MQTT's lightweight nature (small header, simple client implementation) makes it ideal for thousands of sensor nodes. Additionally, MQTT supports features like Quality of Service levels (to ensure delivery) and retained messages (so a new subscriber can immediately get the last value). A significant development for industry is MQTT Sparkplug, a specification on top of MQTT that defines a standard payload format and topic structure for industrial data, ensuring interoperability (so that different vendors' devices can use MQTT with a shared format). While MQTT itself doesn't define the data content, Sparkplug combined with MQTT and a Unified Namespace approach can realise plug-and-play integration in a plant[33]. One might ask, MQTT vs OPC UA? They are complementary: OPC UA focuses on modelling and rich interaction (and can even run over MQTT as a transport), whereas MQTT is a super-efficient pipe for data. Due to its efficiency and scalability, many modern IIoT solutions use MQTT to get data to cloud platforms.

ISA-95 (IEC 62264): ISA-95 is not a communication protocol but an international standard for integrating enterprise and control systems. It provides a reference model for which functions reside at which level (Level 0-1: process/physical, Level 2: control, Level 3: manufacturing operations management/MES, Level 4: enterprise/ERP) and defines the information that flows between these levels. In essence, ISA-95 establishes a common language and structure for data exchange between the plant floor and enterprise systems[28]. For example, it defines objects like Production Schedules, Process Segments, Materials, etc., and how an MES should interface with ERP (often via an intermediate layer sometimes called MOM – Manufacturing Operations Management). The goal is that if both an MES and an ERP adhere to ISA-95 standards (often implemented via B2MML – Business To Manufacturing Markup Language, an XML implementation of ISA-95), they will integrate with minimal custom work. ISA-95 also helps companies map out their architectures clearly, understanding what systems are responsible for what functions and what data should be handed off. It is used as a guiding standard in many requests for proposals to ensure new systems fit into the existing landscape. In recent trends, some propose evolving beyond strict ISA-95 layers to more unified architectures (where data is not strictly siloed by level)[28]. But even then, ISA-95's definitions of data elements remain useful. The standard has also become important in the context of IIoT because it gives a framework to categorise data: e.g., an "Equipment ID" in Level 2 means the same as "Equipment ID" used in Level 4 when following ISA-95, aiding interoperability.

Other Standards and Protocols:

OPC DA / OPC Classic: Predecessor to OPC UA, based on Windows COM/DCOM, still used in older systems for device communication.

- PROFINET, EtherNet/IP, Modbus, CC-Link IE, etc.: Industrial Ethernet protocols that handle real-time control data on the factory floor. They are often below the level of concern for MES/IT but critical for machine-to-machine communication (for example, robotic cell controllers or PLCs using these to coordinate).

- Fieldbus protocols (CAN, PROFIBUS, DeviceNet, etc.): Older serial-based protocols for device networks, gradually replaced by industrial Ethernet variants.

- SQL/ODBC standards: Many machines or devices might log to embedded databases; using standard SQL connectors can be a way to retrieve data if direct IIoT integration isn't available.

- Data Format Standards: Such as JSON, XML, or Automation ML for exchanging data, and CSV remains common for simple file-based transfer. Also, QR codes/barcode standards (like Code128, DataMatrix) are used in the context of traceability for marking and scanning.

- ML Models Exchange: Emerging standards like PMML (Predictive Model Markup Language) or ONNX for exchanging AI models between systems, relevant when deploying ML from cloud to edge in factories.

Standards adoption can vary, but overall, these protocols drastically reduce integration time. For example, a sensor from Vendor A and software from Vendor B can work together out of the box if both support MQTT or OPC UA in a standard way. Similarly, a MES and ERP designed for ISA-95 will share a common understanding of production orders and inventory data. Adhering to open standards also helps future-proof a facility – it mitigates vendor lock-in because you can replace one component with another standard-compliant one without rewriting everything.

In conclusion, protocols like OPC UA and MQTT act as the digital lingua franca on the factory floor, while ISA-95 provides the blueprint for how information should flow. Using these standards is a recognised best practice to achieve the interoperability needed for a successful digital manufacturing strategy.

Future Trends in Manufacturing Data

As manufacturing continues to evolve, several emerging trends and technologies shape how data is used and managed. These trends promise to further increase the intelligence and value extracted from data, and to transform operational models in the industry:

AI and Machine Learning for Deeper Insights

Artificial intelligence (AI) and machine learning (ML) are set to play an even larger role in manufacturing data analysis. While many plants already use ML for specific predictive maintenance or quality models, the future will see more holistic AI-driven systems. This includes advanced neural networks that can analyse images, sounds, and complex sensor fusion to detect patterns that humans or simple algorithms would miss. For example, AI-driven visual inspection using deep learning can identify microscopic defects on a production line far faster and more accurately than manual inspection. ML models can also manage complex multivariate processes. For instance, in chemical manufacturing, AI can control dozens of interrelated parameters to optimise yield and quality simultaneously, something traditional PID loops can't do collectively.

Moreover, AI enables anomaly detection at scale: unsupervised learning can establish baselines of regular operation for each machine and instantly flag any unusual behaviour, even without a pre-defined rule. This adaptive learning system improves over time as it ingests more data, essentially "learning" the nuances of each equipment's behaviour. In maintenance, beyond predicting failures, AI is beginning to prescribe maintenance actions (e.g., ordering a part and scheduling a technician automatically based on the type of failure predicted).

Another frontier is reinforcement learning applied to real-time control – essentially self-optimising feedback loops where an AI "agent" tries adjustments in a simulation (digital twin) and learns the optimal policy to control a process. This could give rise to self-driving factories where many control decisions are made by AI agents that continuously learn and adapt.

AI/ML is also improving supply chain and demand forecasting with better algorithms for scenario planning. Generative AI may come into play in manufacturing by optimising designs for manufacturability or by generating synthetic data to augment real datasets for model training. The key enabler of all these AI applications is the large volume of high-quality data that modern factories are starting to capture – AI thrives on data. We can expect to see more manufacturing companies hiring data scientists and machine learning engineers, or using cloud-based AutoML tools, to leverage their data beyond traditional statistical analysis. The ultimate vision is a factory where AI handles many routine optimisations, and human workers focus on oversight, innovation, and handling exceptions.

Data Monetisation and New Business Models

Manufacturing data is not only valuable for internal improvements – it is increasingly becoming a product or service of its own. Data monetisation refers to extracting economic value from data, either by using it internally to save cost/increase revenue or by selling it externally[34]. Internally, every use case mentioned (optimisation, quality, maintenance) is a form of monetising data (turning data insights into cost savings or higher output). Externally, manufacturers are exploring new business models around data. For example, an equipment manufacturer might offer Analytics as a Service to the buyers of their machines: the machine sends operational data back to the manufacturer, who then provides the customer with insights (benchmarking performance against peers, predictive maintenance alerts, etc.) as a subscription service. This not only adds revenue but also builds customer loyalty. Another example is that companies can create data marketplaces or participate in industry data exchanges where non-sensitive data (such as machine performance and energy efficiency metrics) is shared or sold to others who can aggregate and derive broader insights. Some manufacturers monetise data by partnering with suppliers, sharing production forecasts and inventory data in real time so that suppliers can adjust (often in exchange for better terms or lower prices due to efficiency gains).

Product as a Service model: Instead of selling a machine, a manufacturer might rent it and charge for usage. In this model, data from the machine (like operating hours or cycles) directly drives billing ("X dollars per 1000 parts produced"). This effectively turns data into the meter for revenue. It's seen in models like tire companies charging per mile of tire usage rather than selling tires, using IoT to track it.

The value of data is so high that research suggests manufacturers are heavily investing in data monetisation strategies. For instance, one leadership survey indicated manufacturers in 2023 are investing 9 times more in data monetisation than they did in 2016[35]. Yet, there's also a gap in execution – many executives feel they're not fully realising their data's value[25]. In the future, we'll likely see more Chief Data Officers in manufacturing firms and dedicated data monetisation teams aiming to generate direct ROI from data projects. Data may even appear as an asset on balance sheets if accounting standards evolve.

One challenge in external monetisation is addressing privacy and IP concerns – companies are cautious about what data to share. This leads to the next trend, which helps mitigate that issue.

Federated Learning and Collaboration Across Boundaries

Federated learning is an emerging approach in AI that trains algorithms across multiple decentralised data sources without requiring raw data to be transferred or centralised. In manufacturing, this has enormous potential because often one plant or company alone has limited data on certain rare events, but collectively, there's a lot of data industry-wide. However, companies are reluctant or legally unable to share their raw data due to competition and confidentiality. Federated learning offers a solution: algorithms (models) can be trained collaboratively by sending only model updates, not the raw data, between participants. Each site or company keeps its data local and trains a model on it; only the learned model parameters or insights (which are not sensitive) are shared with a central coordinator that aggregates them into a global model. Thus, all participants can benefit from a much larger effective dataset without exposing their proprietary information.

Imagine a consortium of aerospace manufacturers, each with its machine tool data. They could use federated learning to jointly train a predictive maintenance model for CNC machines that is far more accurate than any single company's model, because it has essentially learned from the collective experience of all. Yet none of the companies had to give their actual data to others – they only exchanged model coefficients. This approach is privacy-preserving and IP-protecting[26], which is crucial in competitive industries.

 Federated learning is still in early adoption in manufacturing. Still, interest is growing in the broader trend of secure data collaboration (also seen in ideas like data trusts and secure multi-party computation). We can expect more platforms and standards to arise that facilitate federated analytics among trusted partners (perhaps led by industry groups or neutral technology providers). This could accelerate innovation by pooling insights – for example, improving quality prediction models by learning from many factories, or enhancing supply chain risk models by sharing demand forecasts without exposing customer specifics.

Self-Optimising and Autonomous Systems

Building on AI and advanced analytics, the vision of the self-optimising factory is becoming a reality. Self-optimising systems can automatically adjust and improve their performance with minimal human intervention. This could mean manufacturing lines that autonomously adapt to variations and disturbances. Key elements include:

Autonomous Process Control: Using AI (potentially reinforcement learning or advanced model-predictive control), processes can fine-tune themselves. For instance, a brewing process might continuously adjust temperatures and ingredient flow based on real-time taste sensor feedback to maintain quality, learning and improving with each batch.

Automated Scheduling and Resource Allocation: Factories will use algorithms to dynamically schedule production, maintenance, and logistics, integrated, reacting to changes like machine outages or rush orders instantly. A self-optimising system could reshuffle production orders in seconds after a breakdown to minimise downtime across the factory.

Robotics and AGV Swarms: Autonomous mobile robots could coordinate among themselves for material handling, dividing tasks optimally without a fixed script, based on current factory state (like a multi-AGV system figuring out the best routes and job assignments on the fly to clear a backlog of material transfers).

End-to-End Supply Chain Adjustment: A future factory might automatically adjust not just internally, but also send signals to suppliers or distribution when it sees a trend (like increasing defect in a supplied part triggers an automated message to the supplier to check their process, or an unexpected surge in demand triggers an immediate order for more raw material, etc., all AI-driven).

A concrete example of the self-optimising trend is what some industry visionaries call the "lights-out factory" – a facility that could run with almost no human intervention, using robotics and AI to handle most scenarios (humans are only needed for oversight and for handling novel situations or making improvements to the system). While truly lights-out operations exist only in limited cases today (for example, some highly automated warehouses or electronic manufacturing at night shifts), the advancements in data and AI are expanding the possibilities.

Digital twins will play a role here too – the self-optimising factory uses its digital twin to test optimisations virtually first (as discussed) and then implement them. Ubiquitous sensing and connectivity are prerequisites – everything needs to be measured and accessible to the AI for a system to optimise itself fully.

Notably, one source describes that self-optimising factories will leverage AI to autonomously adjust production parameters, resource allocation, and maintenance schedules based on real-time data and goals[36]. This succinctly captures the essence: using real-time data, AI will juggle all the variables (machines, people, materials, energy, schedules) to meet objectives (output targets, cost, quality, due dates) without constant human micromanagement. The benefit expected is a new level of efficiency and agility, essentially operating at a state that humans alone could not consistently maintain because of the complexity and speed of decisions required.

 Of course, this is an evolutionary trend – plants will incrementally move toward more autonomy as trust in AI grows and systems demonstrate reliability. Many will start with "self-optimising" specific subprocesses (maybe an AI optimising one machine or one production cell) and then expand outward.

Data Governance and Democratization in Organizations

An important future trend is not a technology but a practice: stronger data governance and data democratisation. As the volume of data grows, companies are recognising the need for formal governance to ensure data is curated, reliable, and used ethically. We will likely see more unified data governance frameworks in manufacturing firms that set policies on data ownership, data retention, and access control. This is crucial not only for data quality, but also for compliance (think of personal data in employee logs or machine vision possibly capturing human faces – privacy laws may require proper handling).

On the flip side, democratisation means making data and analytics accessible to a broader range of roles, not just analysts. Frontline workers might get simplified dashboards or even AR (Augmented Reality) interfaces to see data in context (e.g., a maintenance tech pointing a tablet at a machine to see its recent performance data and predictive alerts immediately). Low-code or no-code analytics tools may allow process engineers (who aren't coding experts) to build their own data queries or simple ML models.

Cloud technologies are also enabling smaller manufacturers to use advanced data analytics without the need for extensive in-house IT by subscribing to platforms or using open-source tools supported in the cloud. This trend levels the playing field, allowing even medium or small enterprises to join the Industry 4.0 movement.

In summary, the future will bring more innovative algorithms (AI everywhere), new ways to collaborate on data (federated learning across companies), new value streams from data (monetisation), more autonomous operations, and a continued emphasis on governing data properly. Manufacturing organisations that stay ahead of these trends will likely be the ones driving innovation and efficiency in the coming decade.

Best Practices for a Data Strategy in Digital Manufacturing

Manufacturers should develop a robust data strategy to navigate the digital transformation successfully. This involves not just technology selection, but also governance, architecture, and cultural elements. Here are the best practices and guiding principles for building an effective manufacturing data strategy:

Establish Strong Data Governance: Data governance is the foundation that ensures all other data initiatives don't collapse. It involves setting up policies, standards, and roles for managing data throughout its lifecycle. For example, define who owns the quality data, who can access machine logs, and what the retention policy is for sensor data. A data governance committee or team can oversee these rules and handle exceptions. As Amplifi succinctly puts it, "Data Governance is the linchpin of any data strategy... take it out, and data quality declines, platforms lose value, and data-driven actions go wrong"[(37)]. In practice, governance includes:

- Data Quality Management: Implement processes to monitor and cleanse data continuously. This could mean automated validation scripts that catch out-of-range values, or periodic audits comparing data across systems for consistency.

- Master Data Management (MDM): Ensure key identifiers (equipment IDs, product codes, customer IDs) are consistent across all systems. This might involve maintaining a master database that all systems reference, to avoid, say, one system calling a product "123-XL" and another "123XL" and them not matching.

- Metadata and Cataloguing: Use data catalogue tools to document what data exists, where, and in what format. This makes it easier for analysts or engineers to find the data they need and understand its context.

- Security and Privacy Policies: Define access controls for each data category. Adopt the principle of least privilege (people/systems only access what they need). Also, ensure compliance with privacy regulations or customer agreements regarding data (for instance, if you gather data from a client's equipment, how will it be used?).

- Data Lifecycle and Retention: Not all data needs to be kept forever. Define retention schedules to delete or archive data that is no longer useful, to reduce storage costs and liability. Some regulations might require minimum retention (e.g. keep batch genealogy for X years in pharma).

Having governance builds trust in the data; users are more likely to use data if they know it's been appropriately validated and is accurate. It also aligns the organisation with the importance of data as an asset.

Design a Scalable, Integrated Data Architecture: The architecture is how all your data sources, pipelines, and storage fit together. A best practice is to create a unified architecture where data from different systems can come together, rather than isolated point solutions. For instance:

- Use a Central Data Platform or Lake: Consolidate copies of critical data into a central repository (cloud or on-prem) that can be the hub for analysis. This could be a data lake that stores raw data and a data warehouse that stores curated, query-optimised data. The central platform should be able to pull from MES, ERP, historians, etc., using connectors or APIs in real-time or through regular polling. This doesn't necessarily replace those source systems (they continue doing their job), but it provides a one-stop shop for analytics and reporting. Many companies choose cloud platforms (like AWS, Azure, GCP) for this due to their scalable services.

- Adopt Standard Protocols and Interfaces: When connecting machines and systems, prefer standard protocols (OPC UA for machine data, RESTful APIs or MQTT for IoT, SQL interfaces for databases, etc.) instead of proprietary ones. This ensures each new system can plug in with less custom work and that you can swap out components. Also, consider using an enterprise service bus (ESB) or IIoT integration platform that handles message translation and routing between systems.

- Edge and Cloud Balance: Implement an edge architecture for low-latency needs and a cloud/central architecture for heavy analytics. Define clearly what data should be processed at the edge (e.g. immediate control decisions, data reduction tasks) and what should be sent upstream. Use buffering and store-and-forward techniques at the edge to prevent data loss if connectivity is intermittent.

- Unified Namespace or Data Model: As a more advanced practice, some organisations implement a unified namespace – essentially a structured hierarchy (often implemented via MQTT topics or similar) that mirrors the ISA-95 hierarchy or their logical model of the plant. All data is published into this namespace. For example, Enterprise/Plant/Area/Line/Machine/DataType. This consistent organisation makes it easier to integrate new devices (they publish to a known part of the namespace) and for consumers to discover relevant data.

- Modular, Layered Design: Structure your architecture in layers (sensing/actuation layer, network layer, data ingestion layer, storage layer, analysis layer, application layer), so you can tackle and upgrade each layer somewhat independently. For instance, if the layers are separated with standard interfaces, you might replace your visualisation tool (application layer) without re-engineering how data is collected from machines (ingestion layer).

The architecture should also be scalable, as data volumes are expected to grow as you add more sensors and higher resolution data. Cloud elasticity, distributed computing, and scalable databases (like time-series DBs that can handle millions of events) are key. Testing the system under load and planning capacity ahead of needs is part of best practice.

Invest in Analytics Infrastructure and Tools: Having data is one thing; having the tools to extract value is another. Best practices include:

- Advanced Analytics and BI Platforms: Deploy user-friendly analytics tools for different levels of users. For business users or operations managers, BI dashboards (Tableau, Power BI, etc.) that pull from the data warehouse might suffice for descriptive analytics. For data scientists, provide access to notebooks (Jupyter, RStudio) and big data processing (Spark clusters, SQL engines, etc.) on the data lake. Ensure tools can handle real-time data where needed (some dashboards might need to show data with a latency of seconds).

- Machine Learning Infrastructure: Set up an environment for developing and deploying machine learning models. This could involve cloud ML services or on-prem GPU servers for training models on historical data, and then edge or cloud deployment for inference (predictions) in real time. The infrastructure might include automated data pipelines for training (so models can be retrained with new data periodically) and monitoring for model performance drift.

- Visualisation and Alerting: Besides formal dashboards, consider large shop floor displays showing real-time KPIs to workers, encouraging a data-driven mindset at all levels. Implement alerting systems (which could be part of MES or a separate system) that send notifications (email/SMS/alarms) when certain data conditions occur (machine down, quality issue emerging, etc.). The threshold and rules for alerts should be fine-tuned to avoid alarm fatigue.

- Digital Twin Simulators: If applicable, invest in simulation tools that can serve as digital twins for either process optimisation or scenario planning. This might involve using existing software (like process simulation for chemical plants, or discrete event simulation for production lines) and integrating it with live data. Over time, this can merge with analytics infrastructure, so that what-if scenarios can be run using authentic data snapshots from the plant.

- Historian Upgrades: If the facility does not have a modern historian or if it's siloed, consider upgrading or linking it to the enterprise data pool. Modern historians can forward data to the cloud or integrate via OPC UA easily, acting as a strong bridge between OT and IT.

- Edge analytics: In some cases, the edge devices are equipped not just with data forwarding capabilities but also with analytics (for example, an AI gateway that does vibration analysis on the fly and only sends anomalies). This requires tools to deploy models to the edge (like Azure IoT Edge, AWS Greengrass, or other containerization) and to update them remotely.

Choosing the right tools also means evaluating where to buy vs build. Many vendors offer integrated solutions (for example, IoT platforms that include connectivity, storage, and analytics). These can accelerate deployment, but beware of vendor lock-in. It's often best to choose tools that adhere to open standards and that can plug into your broader architecture.

Develop Human Talent and a Data-Driven Culture: Technology alone doesn't create value; people do. Fostering an organisational culture that values data is perhaps the most critical success factor. Key practices:

- Executive Sponsorship and Vision: Leadership should clearly articulate the importance of data in the company's strategy. When the C-suite uses data in their decisions and talks about improvements quantitatively, it sets the tone. Investment in data initiatives should be sustained and protected by leadership, even if early experiments fail or take time to pay off.

- Cross-Functional Teams: Break down the wall between IT and OT teams by creating cross-functional groups for data projects. For example, a predictive maintenance project team might include a data scientist, a maintenance engineer, a control systems engineer, and an IT cloud architect. This ensures solutions are practical and get buy-in from end-users. It also facilitates skill transfer; the engineer learns some data science, and the data scientist learns some domain knowledge.

- Training and Upskilling: Invest in training programs to improve data literacy at all levels. This could mean teaching shop floor supervisors how to interpret control charts, training engineers on using SQL or Python for data analysis, or upskilling IT folks on OT protocols. Some companies establish an internal "data academy" or use online courses to elevate skill levels. When introducing new tools (say a self-service analytics tool), provide hands-on workshops using actual company data so employees see the relevance.

- Empowerment and Incentives: Encourage employees to use data in problem-solving by giving them access to data (with appropriate governance) and tools. Celebrate wins where data analysis led to cost savings or solved a quality problem. Some firms run data hackathons or kaizen events where teams compete or collaborate to analyse data and find improvements, fostering a bit of competition and fun around data. Align incentives so that managers reward data-driven decision making rather than just "gut feeling" heroics.

- Change Management: Recognise that moving to a data-driven operation is a change that might face resistance. Engage with those who are sceptical; show them how data tools can make their jobs easier rather than threatening their expertise. For example, experienced operators have a wealth of knowledge, so they involve themselves in training AI models by labelling data or validating algorithm recommendations, which they view as a way to capture their knowledge, not replace it.

- Communication: Keep the organisation informed about data initiatives. Share success stories: e.g., "Using our new analytics system, the team in Plant A reduced defect rate by 20% last quarter." This not only motivates but also provides concrete examples to learn from.

Start with Clear Use Cases and Iterate: Best practice is to begin your data strategy with a few well-defined, high-value use cases (like predictive maintenance on critical equipment, or yield improvement on a bottleneck process) rather than trying to "boil the ocean" all at once. Delivering a successful project builds momentum and justifies further investment. Use agile methodologies – quickly develop a minimum viable product (MVP) for the use case (maybe a basic dashboard or a pilot predictive model), then iterate with user feedback. This helps refine requirements and trains the organisation on how to work with data. Over time, expand to more use cases and gradually integrate them (for example, once you have several successful analytics applications, you might incorporate them into a larger digital twin or analytics portal).

Ensure Interoperability and Avoid Vendor Lock-In: When implementing systems (MES, IoT platforms, etc.), favour those that support open standards and easy data export. Proprietary data formats or systems that don't play well with others can create dead ends in your data strategy. Getting a monolithic solution from one vendor may be tempting, but if that solution is closed, it will hinder innovation. A mix-and-match ecosystem with open interfaces often gives more flexibility to adopt new tech. Using standards and protocols (as discussed earlier: OPC UA, MQTT, ISA-95, etc.) is a key practice here. Also, maintain documentation of your integrations and data models, so the knowledge isn't lost if you switch vendors or upgrade parts of the system.

Focus on Cybersecurity from Day One: Incorporate security into the design of your data architecture (often called "security by design"). This includes network segmentation (keep critical control networks isolated with secure gateways to the IT network), using VPNs or secure tunnels for any remote access, regular patches of software (with testing to ensure it doesn't disrupt operations), and possibly deploying industrial-specific security solutions (like intrusion detection for SCADA protocols). Educate employees on cybersecurity hygiene, as even the best system can be compromised via phishing or a USB drive. Frequent backups of critical data (and plans for restoring operations without complete data in worst-case scenarios) are also part of a resilient strategy – ransomware has hit many manufacturers, and having backups and fallback procedures can make the difference in recovery.

Measure and Refine: Finally, treat the data strategy itself as data-driven. Define KPIs for your digital initiatives – e.g., percentage of data sources integrated, reduction in decision lead time, cost savings achieved from analytics, user adoption rates of new tools, etc. Track these over time. Solicit feedback from users on what's working and what pain points remain (perhaps the UI of a tool is not user-friendly, or a particular report is frequently inaccurate – fix those quickly). The digital journey is continuous; technology and best practices evolve, so keep updating the strategy. For instance, what's state-of-the-art today (maybe cloud analytics) could be augmented tomorrow by new tech (like edge

AI chips on every device). A learning organisation will keep pace by staying flexible and watching new developments.

In essence, a data strategy in manufacturing should align with the company's overall business strategy (e.g., if the goal is to be the quality leader, focus data efforts on quality analytics and traceability; if it's operational efficiency, focus on optimisation and maintenance analytics, and so on). It requires the right mix of people, process, and technology. Companies that successfully implement these best practices are positioned to exploit their data as a competitive asset, improving responsiveness, reducing costs, innovating faster, and even creating new revenue streams in the era of digital manufacturing. As a closing thought: data should be treated with the same rigour as any physical asset on the shop floor. Just like machines need maintenance and skilled operators, data requires governance and skilled analysts; when the physical and digital assets are finely tuned and working in concert, the modern manufacturing enterprise can reach new heights of performance[25, 37].

Robotics and Automation in Digital Manufacturing

Robotics and automation have become cornerstones of modern digital manufacturing, transforming how products are made in industries from automotive to electronics. This section comprehensively overviews how various robots and automation systems contribute to smart manufacturing. It covers the main types of industrial robots and their functions, the enabling role of automation technologies like PLCs, machine vision, and AI, as well as the benefits and challenges of adopting robotics. We also discuss how robots integrate with digital manufacturing systems (MES, PLM, ERP, digital twins), real-world industry examples, current trends shaping the future (AI-driven adaptive automation, human-robot collaboration, cloud robotics, and key standards like OPC UA and ISA-95), major robotics vendors, and best practices for planning, deploying, and scaling robotics in production. Manufacturing professionals will gain a broad perspective and technical detail on leveraging robotics for operational excellence.

Types of Robots in Manufacturing

Industrial robots come in many forms, each suited to different tasks. Industrial robots are generally defined as "automatically controlled, reprogrammable, multipurpose manipulators programmable in three or more axes"[(38)]. They can be fixed or mobile; today, a wide variety exists to meet specific manufacturing needs. The table below summarises the major robot types commonly used in factories:

Robot Type	Structure & Degrees of Freedom	Typical Use Cases
Articulated Robot	Rotary joints (usually 4–6 axes) allow a wide	Versatile tasks: welding, painting, assembly,

	range of motion, similar to a human arm[39].	machine tending, material handling[39].
SCARA Robot	4-axis robot (Selective Compliance Assembly Robot Arm) with two parallel rotary joints for horizontal movement, plus a vertical axis[39].	High-speed lateral movements for assembly, pick-and-place, palletising, and electronics manufacturing[39].
Delta Robot	A parallel-link robot with three lightweight arms is connected to a common base, creating a dome-shaped working envelope[39]. Typically, 3-4 axes, including an end-effector orientation.	High-speed pick-and-place operations in food, pharmaceutical, and electronics packaging where precision and speed are critical[39].
Collaborative Robot (Cobot)	Various forms (often a 6-axis articulated arm) are designed with force-limited joints, vision, and safety sensors to work alongside humans[39].	Working in close proximity to humans for light assembly, machine tending, quality inspection, and packaging, without safety cages (fenceless operation)[39].
Mobile Robot (AGV/AMR)	Wheeled robots for material transport. AGVs (Automated Guided Vehicles) follow fixed paths (tracks, wires, or markers) and require a structured environment[40]. AMRs (Autonomous Mobile	Intralogistics tasks: moving parts and materials across the factory or warehouse. AGVs excel at repetitive transport on predetermined routes (e.g., assembly line supply), while AMRs offer flexible routing in dynamic

	Robots) use onboard sensors, cameras, and AI for self-navigation without fixed infrastructure[40].	warehouse environments [40].

Table: Type of robots and their use cases

Articulated robots are the most common industrial robots; they have multiple rotary joints and often a six-axis configuration for whole 3D movement[39]. These robots are highly flexible – a single articulated arm can weld a car frame, pick and place components, or palletise boxes by simply changing its tooling and program. SCARA robots (a subset of articulated designs) have a more limited motion (usually four axes) but are extremely fast and precise in horizontal movements[39]. SCARAs are widely used for assembly operations like PCB component placement or small parts insertion, where speed and repeatability are paramount. Delta robots are known for their spider-like parallel linkages, allowing them to achieve very high accelerations and precise movements for lightweight payloads[39]. They are commonly suspended above conveyor lines to perform lightning-fast picking, sorting, or packaging tasks (for example, sorting candies or placing cookies into trays) in industries that demand both hygiene and speed.

A newer category, collaborative robots (cobots), is designed for direct interaction with human workers. Unlike traditional robots that usually operate in fenced-off cells, cobots have built-in safety features (such as torque sensors to detect collisions, safe speed limits, and soft end-of-arm tooling) that enable them to work side-by-side with people[39]. This opens up applications where robots assist humans rather than fully replace them – for instance, a cobot might hold a part steady. At the same time, a technician performs an assembly step, or it might take over repetitive tasks on a production line while humans handle more complex tasks. Cobots typically have user-friendly programming (often via handheld teach pendants or even direct hand-guiding for teaching motions). They can be redeployed easily, which appeals to small and medium manufacturers without large automation engineering teams[41].

Another primary class are mobile robots that move materials around the facility. Traditional automated guided vehicles (AGVs) are essentially robotic carts or forklifts that follow fixed pathways (tracks embedded in the floor, magnetic tape, or laser-guidance) and are used to transport goods within factories or warehouses. They've been used for decades in automotive plants for line-side delivery of parts, or in distribution centres to shuttle pallets. AGVs, however, lack flexibility – they cannot deviate from their guide path and will stop if an unexpected obstacle blocks the route. In contrast, autonomous mobile robots (AMRs) represent the next generation of mobile robotics: they navigate using onboard intelligence (lidar sensors, 3D cameras, SLAM algorithms) to dynamically plan paths and avoid obstacles in real time[40]https://www.mwes.com/agvs-vs-amrs-whats-the-difference-and-which-one-is-right-for-you/. AMRs don't need predefined routes or physical guides; this makes them more adaptable to factory layouts or workflow changes. They are increasingly popular in e-commerce warehouses and manufacturing, addressing material handling needs with greater flexibility – for example, AMRs can dispatch parts between workstations on demand or handle order picking in warehouses, working safely around human workers. AGVs and AMRs are part of the broader trend of mobile robotics that extends automation beyond fixed production stations to the entire facility's logistics.

In addition to the above, other robot architectures exist (though are somewhat less prominent today). Cartesian or Gantry robots move in straight lines along X, Y, Z axes (and sometimes a rotary axis)[39]; these are essentially CNC machine or 3D-printer style robots and are excellent for heavy payloads and simple point-to-point tasks like pick-and-place over a large area. Cylindrical and polar (spherical) robots were early industrial robot designs with a mix of linear and rotary joints that give a cylindrical or spherical work envelope[39]. While not as common now, they historically found use in tasks like simple assembly or machine loading/unloading in constrained spaces. These additional types illustrate that robot kinematics can be tailored to the task geometry. Still, the articulated arms, SCARAs, deltas, cobots, and mobile robots are the primary workhorses of modern digital factories.

Automation Systems Enabling Smart Manufacturing

Robots seldom work alone – they are part of a larger automation system that includes various controllers, sensors, and software orchestrating manufacturing operations. Key components include programmable logic controllers (PLCs), machine vision systems, and increasingly AI-driven control systems (often tied into Industrial IoT platforms). These technologies act as the "brains and senses" of an automated facility, enabling smart manufacturing where equipment can make decisions, adapt, and coordinate in real time.

Programmable Logic Controllers (PLCs) and Industrial Control: PLCs are rugged industrial computers designed to run automation routines on the factory floor reliably. Since their introduction in the 1970s, PLCs have been the backbone of factory automation, replacing complex relay logic with flexible software logic. PLCs monitor sensors and machine status, execute programmed logic (e.g. sequencing steps, safety interlocks), and send commands to actuators like robot motors, conveyors, valves, etc. They operate in real time and are highly reliable. PLCs enabled the third industrial revolution by allowing automated machinery control with minimal human intervention in repetitive tasks[42]. In a robotics context, PLCs often coordinate multiple devices. For example, a PLC might supervise an entire robotic cell, handling communication between the robot arm, part feeders, safety scanners, and other machinery. PLCs can start or stop robot programs, handle emergency stop signals, and synchronise robot motions with different equipment (like ensuring a press is open before a robot inserts a part). Modern PLCs are networked and can feed data to higher-level systems, bridging shop-floor devices and supervisory software.

Machine Vision Systems: Vision technology allows automated systems to see and interpret visual information. Machine vision cameras (often paired with AI algorithms) are used for quality inspection, part identification, guidance, and measurement in manufacturing. For example, a vision system can detect if a part has defects or if components are correctly assembled, and then signal a robot to reject or rework the item. Vision-guided robots use cameras to locate the position of parts so the robot can pick them up even if they're not in exact known locations (this is crucial for tasks like bin picking, random parts, or picking items off a moving conveyor). Artificial intelligence (AI), particularly deep learning, has dramatically improved the capabilities of vision systems – AI-powered cameras can detect subtle defects or classify products with high accuracy, far beyond simple rule-based image processing[42]. In smart factories, machine vision is often integrated at multiple stages: checking incoming materials, monitoring production in-process (for example, detecting a misaligned part and alerting the robot or PLC to correct it), and performing final quality control on finished goods. The data from vision inspections can be fed into databases to track quality trends and trigger adjustments. Vision systems thus act like the quality control eyes of an automated line, working at high speed and with consistency that manual inspection can rarely achieve.

AI-Driven Control and Industrial IoT: Beyond fixed programs, manufacturers are increasingly infusing artificial intelligence and machine learning into automation for greater adaptability and optimisation. AI-driven control systems can analyse sensor data from machines and robots to optimise real-time parameters or predict and prevent problems. For example, AI algorithms can monitor the electrical signals of a robot's motors and predict an impending maintenance need (predictive maintenance), avoiding unplanned downtime. Advanced robotics today often combines sensors and AI to handle complex tasks. For instance, force-torque sensors on a robot combined with AI allow it to "feel" its way into fitting a part accurately, or to adjust force when polishing a surface. According to industry reports, AI and machine learning significantly enhance manufacturing productivity and decision-making quality[42]. Robots are becoming more intelligent agents on the shop floor: they can optimise their paths, detect anomalies in their operation, and even learn new tasks through machine learning techniques. A notable example is vision-guided robotics: AI allows robots to see and make judgment calls (e.g., identifying the best grasp points for irregular objects on the fly, or adapting to variable input). The

Industrial Internet of Things (IIoT) further supports AI-driven automation by connecting robots, machines, and sensors into a network where data is continuously collected and analysed [42]. This connectivity means a robot's data can be sent to a cloud platform where advanced analytics or AI models compute optimisations and send back recommendations or control inputs. In sum, AI-driven control and IoT connectivity are making automation more responsive and smart, moving from strictly pre-programmed sequences to adaptive systems that can adjust to changing conditions in a manufacturing environment.

By combining robust PLC control, the sensory insight of machine vision, and the adaptive intelligence of AI, manufacturers realise the vision of smart manufacturing (often dubbed Industry 4.0). In smart factories, robots and other machines are not isolated – they communicate, coordinate, and adjust continuously, leading to higher efficiency and quality. For example, if a vision system detects a trend of slight misalignment in an assembly, an AI controller could instruct a robot to slightly change its path or force at specific steps, or alert a human supervisor if intervention is needed. This tight integration of robotics with advanced control technologies differentiates modern digital manufacturing from the past's automated but more rigid production lines.

Benefits of Robotics and Automation

Implementing robotics and automation in manufacturing offers a multitude of benefits. When designed and deployed well, robots can dramatically improve productivity, quality, and safety, while providing new flexibility in operations. Some of the key advantages include:

Higher Productivity and Consistent Quality: Robots work rapidly and tirelessly. They can operate 24/7 without fatigue, maintaining a high throughput that would be impossible with manual labour alone. This boosts output rates and can shorten production lead times. Equally important, robots perform tasks with precision and repeatability, ensuring consistent quality. A well-programmed robot will perform the same task the same way every cycle, minimising the variability that can come with human work. This consistency reduces errors and scrap. In many cases, introducing robotic automation has increased production output while lowering operating costs due to fewer mistakes and less waste. For example, robotic arms can produce uniform welds or coats in welding or painting, improving product durability and finish. Over time, the combination of higher speed and quality translates to a higher return on investment (ROI) for the manufacturer[43].

Labour Cost Reduction and Addressing Labour Shortages: Although the upfront investment in robots is significant, they can reduce direct labour costs in the long run. One robot can often do the work of multiple human shifts for specific repetitive tasks. Moreover, many industries face a shortage of skilled labour – fewer people are interested in doing repetitive factory work, and demographic trends in some countries mean there aren't enough workers. Automation helps fill this gap. By automating the "dull, dirty, and dangerous" jobs, companies can allocate their human workforce to more value-added roles. As one industry analysis noted, robots are often adopted to replace a lack of available workforce, especially as younger generations seek different kinds of jobs[43]. This was seen in recent years, where record-low unemployment and an ageing workforce pushed manufacturers to accelerate robot adoption to meet demand. Additionally, robots can help avoid production slowdowns due to absenteeism or shift changes – they provide very reliable production capacity.

Enhanced Flexibility and Agile Production: Modern automation is far more flexible than the complex automation of the past. Reprogrammable robots allow manufacturers to change over production lines for new products or variations rapidly. This is crucial as product lifecycles shorten and customers demand more customised products. For instance, a robotic cell can be reprogrammed in hours or days to handle a new model or a different task, whereas retooling a fully manual line might take much longer. This flexibility supports high-mix, low-volume production. Collaborative robots, especially, are touted for quick re-deployability – they often come with easy programming interfaces and modular end-effectors (grippers, tools) that can be swapped for new tasks. As a result, even smaller manufacturers can stay competitive by producing customised orders without huge reconfiguration costs[43]. Automation also enhances scalability: once a process is automated, ramping up production (by adding more robots or increasing robot working hours) is more straightforward and more linear than hiring and training many new workers for a sudden demand spike.

Improved Worker Safety and Ergonomics: One of the most humane benefits of robotics is taking over dangerous or ergonomically harmful tasks from people. Robots excel at tasks that involve heavy lifting, exposure to hazardous substances, or repetitive motions that can cause strain injuries. Robots can handle hot, toxic, or sharp materials in industries like metals or chemicals, keeping humans out of harm's way[43]. This not only protects workers from accidents (e.g., robots can work in paint booths with toxic fumes or foundries with molten metal) but also reduces long-term health issues from repetitive or strenuous labour. Automation contributes to a safer work environment and can lower injury rates. Many companies adopting robots report improved worker morale – employees are relieved from drudgery and can focus on supervision, maintenance, or creative problem-solving, which are less physically taxing. Robots also enforce consistent safety behaviour (they won't take unsafe shortcuts), enhancing workplace safety. It's important to note that collaborative robots are specifically designed with safety in mind, as they directly interact with humans; features like power-and-force limiting mean they can sense an obstruction (like a person's arm) and stop before harm[41]. Thus, automation can eliminate the 4 D's of undesirable work – dull, dirty, dangerous, or complex tasks – allowing humans to work in more pleasant and meaningful roles[44].

Higher Process Efficiency and Analytics: Automated systems generate detailed data on operations – cycle times, downtime causes, error rates, etc. This data can be fed into manufacturing analytics or MES systems to identify bottlenecks and optimise processes continuously. For example, engineers might find a way to remove a minor delay and increase throughput by analysing robot cycle data. Automation lends itself to lean manufacturing improvements because of its consistency; any deviation is easier to detect and trace to the root cause. Additionally, robots often enable the reduction of material waste. For instance, automated dispensing systems apply just the right amount of adhesive every time, whereas manual processes might use too much (waste) or too little (quality issue). The precision of robotics thus means less scrap and rework. In packaging, a robot can optimise how it stacks items to minimise wasted space. All these incremental gains add up to significant efficiency improvements. Some studies have shown double-digit percentage improvements in overall equipment effectiveness (OEE) after introducing robotics, thanks to greater uptime and output quality[43].

The benefits of robotics and automation help manufacturers meet higher production demands while staying competitive in cost and quality. These technologies are also crucial enablers for lights-out manufacturing (fully automated production with little to no human presence) in specific contexts, which can further drive efficiency. It should be noted, however, that achieving these benefits requires proper implementation and alignment of the technology with business goals – a theme we will revisit in best practices.

(The benefits above are well documented across industries – from pharma to automotive. For instance, pharmaceutical production has improved quality and sterility by using robotics for filling and inspecting drug vials, minimising human contamination risks[43]. In automotive, welding robots' combination of speed and precision delivers high throughput and strong, consistent weld quality unattainable by manual welding. These examples illustrate why industrial robots have proven valuable in various applications[43].)

Challenges and Limitations

While the advantages are compelling, robotics and automation are not without challenges. Companies must navigate several limitations and obstacles when implementing these technologies in a manufacturing setting:

High Initial Costs: The investment cost for industrial robots and automation infrastructure is substantial. Purchasing a robot (an AMR, or a vision system), the necessary peripherals (end-effectors, safety equipment, controllers), and installation can require significant capital expenditure. There are also engineering costs for custom automation for system integration and programming. As a result, the upfront cost barrier is high, especially for small and medium-sized enterprises. Even though robots often pay off over time through labour savings and efficiency gains, the payback period can be a concern. Additionally, if modifications are needed (for example, reconfiguring a robot cell for a different product), further investment might be required if the original system was not designed with flexibility in mind[43]. Manufacturers must carefully calculate ROI and may need to start with pilot projects to justify the cost. Prices of robots have been gradually declining and financing options (like robot-as-a-service leasing models) are emerging, but cost remains a top consideration.

Integration Complexity: Introducing robots into existing production lines is not a simple plug-and-play affair. It involves integrating hardware, software, and processes. New automated equipment must be compatible with legacy machines and control systems. Ensuring that a robot can communicate with a factory's PLCs, MES, or safety systems requires an integration effort. Each robot needs a well-designed workcell: fixtures to hold parts, sensors to verify positions, safety scanners or fencing, etc. This system integration is often complex and may require the expertise of specialised robotics system integrators. Poor integration can lead to unreliable performance or safety risks. Moreover, robotics interfacing with other digital systems is tricky – data formats, protocols, and timing must all align. It's crucial to select vendors and equipment that adhere to standards and can "talk" to each other[38]. Companies must carefully plan how a new robot will physically and digitally mesh with their production line. Sometimes modifications to the workflow or layout are needed. The complexity of integration can also extend the deployment timeline; installing a robot might only take days, but fully integrating it into smooth operation might take weeks or months of debugging and tuning.

Workforce Skills and Reskilling: Robots don't eliminate the need for human talent – they change it. A workforce used to manual operations needs training to work effectively with automation. There is a learning curve to program and maintain robots and automation systems. Skilled robotics engineers or technicians are in high demand, and not every factory has them on staff. This means companies may need to invest in training existing employees or hiring new talent with automation expertise[43]. For example, maintenance technicians must learn to troubleshoot robot controllers or fix sensor issues, and operators might need to learn how to jog a robot or recover it from an error state. Workforce reskilling is not only a technical challenge but also a change management issue – some workers may be resistant, fearing job loss or feeling anxious about new technology. Successful automation projects often involve transparently communicating the intent (e.g., to make jobs safer, not to replace everyone) and providing robust training programs. Hence, employees feel confident in new roles (such as robot supervisors or automation technicians). A company might under-utilise the robot without sufficient expertise or suffer extended downtimes. Hence, human capital investment must go hand-in-hand with capital investment in robotics[43].

Maintenance and Reliability: Automated systems require regular maintenance to run smoothly. Robots have mechanical components (gears, belts, motors) that wear over time and need preventive maintenance. They also rely on sensors and software that must stay calibrated and updated. A robot breaking down unexpectedly can cause significant production downtime, especially if the process has no manual backup. The cost of unplanned downtime can be higher in an automated line because the line might be designed around the robot with no easy manual workaround[43]. Thus, companies must be prepared with maintenance plans: keeping spare parts, scheduling periodic servicing, and possibly using condition monitoring to predict failures. Some newer robots employ diagnostics and even AI for self-monitoring, but it's not foolproof. Compared to some manual processes, automation can appear less forgiving – a single sensor failure could halt an entire cell. Over-reliance on automation without proper maintenance plans can backfire. Additionally, when a process changes or new products are introduced, automation systems may need re-engineering or re-validation, which is another form of maintenance (software maintenance). Manufacturers should plan for the total lifecycle costs, including ongoing support and eventual upgrades. Collaborating with vendors for service contracts or having trained in-house maintenance staff is essential to keep robotic systems reliable.

Cybersecurity Risks: As factories become digitally connected (with robots, PLCs, and other networked machines, often linked to enterprise IT or cloud systems), they face cybersecurity threats. Industrial robots historically operated in isolated cells, but now that they exchange data with MES or are monitored remotely, they become potential targets for cyber attacks. A hacked robot or controller could lead to data theft and physical consequences. For example, a cyber intruder might manipulate a robot's motion, causing it to damage equipment or endanger workers[45]. There have been demonstrations of how industrial robots or AGVs could be maliciously controlled if not adequately secured, leading to scenarios like production sabotage or even using a robot to cause accidents. Safeguarding industrial automation is thus a new challenge. It involves updating legacy systems that may not have been designed with security in mind (many use outdated operating systems or default passwords). Companies must implement strong network segmentation, firewalls, encryption of communications (e.g., using OPC UA security features), and regular patching of robot and PLC software, which is easier said than done, since stopping a production line for updates is difficult. Standards like IEC 62443 address industrial control system security, and best practices include conducting cybersecurity risk assessments for robotic systems[45]. This is a growing concern: as one industry source noted, increasingly connected and intelligent robots expand the attack surface and require a proactive approach to ensure safe and secure operation[45]. Failing to do so can undermine all the other benefits if a cyber incident disrupts operations.

Other Limitations: There are some additional limitations worth noting. Not every task is easily automated – highly complex processes that require extreme dexterity or on-the-spot decision-making (especially in unstructured environments) might still be better done by humans or require very advanced (and expensive) automation. In some cases, trying to automate a task can even reduce flexibility if not done thoughtfully (a robot might handle a narrow range of product variants unless programmed otherwise). There is also a space issue – robots and safety areas can take up significant floor space, which can be an issue in cramped facilities. Furthermore, the human element cannot be ignored: employee acceptance and adapting the work organisation to human-robot collaboration is non-trivial. Lastly, on the business side, calculating the true ROI of automation can be complex, and there is a risk of underestimating the effort needed to support the automated system. These factors mean that adopting robotics requires a strategic approach rather than a blind rush to automate everything.

Despite these challenges, the trajectory of technology is steadily lowering barriers – costs are gradually coming down, integrators and solution providers are more prevalent, and user-friendly robots address the skills gap. Understanding these limitations allows companies to plan mitigations (like phased implementation, training programs, and cybersecurity measures) and thus reap the benefits of automation while managing risks.

Integration with MES, PLM, ERP, and Digital Twins

Modern manufacturing doesn't view robots as standalone islands of automation – they are integrated into a broader digital ecosystem that includes systems like MES (Manufacturing Execution Systems), PLM (Product Lifecycle Management), ERP (Enterprise Resource Planning), and emerging tools like digital twins. Seamless integration ensures that robotic operations are synchronised with the rest of the enterprise, data flows freely for analysis, and the production process is connected end-to-end from design to delivery. This section explores how robotics ties into these systems:

Manufacturing Execution Systems (MES): MES software sits between real-time control and higher-level business systems, overseeing production execution on the factory floor. When robots are integrated with an MES, the MES can dispatch work orders to robotic cells, monitor their progress, and adjust schedules on the fly. For example, an MES might instruct a robot assembly cell on the product variant to build for the next unit, based on customer orders fetched from the ERP. As the robot works, it can send status updates, cycle times, and any anomalies back to the MES. The MES aggregates data from robots and other machines to give supervisors a real-time view of production. Integration means robots become part of the digital workflow – if one robot goes down, the MES can reschedule tasks to another cell or alert maintenance immediately. MES can also handle traceability: recording which robot processed which part at what time is essential for quality control and compliance (especially in regulated industries like aerospace or pharma). A key part of integration is having standardised interfaces. Many robots and PLCs today support protocols (like OPC UA or MQTT) that MES systems can use to gather data. A well-integrated MES-robot setup allows tight coordination – for instance, a MES might trigger a quality check after every 100 parts a robot produces, or it might instruct an AMR to deliver raw material to a robot cell exactly when needed, based on production rate. One source notes that vertically integrating MES with shop-floor devices provides a real-time view of operations and helps make effective decisions. In practice, custom middleware or IoT platforms are sometimes used to bridge older robot controllers with modern MES, but the end goal is the same: a unified, intelligent production control.

Product Lifecycle Management (PLM) and Digital Thread: PLM systems manage product design data, revisions, and the overall lifecycle from concept through engineering to manufacturing and service. Integration of PLM with robotics occurs through the digital thread, ensuring that the product and process design are linked. For example, when a product design is updated (say, a part geometry changes), the PLM system can propagate that information to manufacturing process planning, including updating the robot's program or tooling to accommodate the new design. Some advanced implementations use virtual commissioning: the robot's programs are prepared and tested in a simulation (a digital twin of the robot cell, which we'll discuss shortly) using data from PLM/CAD models to verify that the new product can be manufactured with the existing robot. PLM can also store process recipes and assembly instructions that MES or robots execute. In essence, PLM integration ensures that the what and how of manufacturing stay in sync – the robots build according to the latest design specifications. This reduces errors from outdated instructions and speeds up changeovers when designs change. It also supports mass customisation: by tying product configurations in PLM to automated setup in manufacturing, a wide variety of product options can be handled without manual intervention. The digital thread connecting PLM (design) -> MES (execution) -> automation (robots) is a hallmark of digital manufacturing, providing traceability and agility.

Enterprise Resource Planning (ERP): ERP systems manage the business side – orders, inventory, supply chain, and financials. Integrating robotics with ERP is typically indirect (often via MES), but important. For instance, when an order is received in the ERP, it may trigger the MES to schedule production, thereby getting robots working on that order. Conversely, as robots complete products, they can inform ERP through MES that inventory is updated or an order is finished. In highly automated warehouses (like automated storage and retrieval systems or robotic fulfilment centres), the WMS (warehouse management system) or ERP might directly command AGVs/AMRs to move goods according to inventory needs. The benefit of integration here is end-to-end visibility: management can see in the ERP that a certain number of units have been produced (because the robots reported it), enabling more accurate shipping commitments or inventory control. Additionally, ERP can feed robots through MES with the needed data, like which recipe to run for a given product variant (since ERP holds the customer order details). This vertical integration from the enterprise level (ERP) to control (robots/PLC) is often conceptualised by the ISA-95 standard, which defines distinct layers and how information should flow between them. In ISA-95, Level 4 (ERP, business planning) connects to Level 3 (MES, operations management), which in turn connects to Level 2/1 (control and devices, including robots)[46]. By following this framework, companies ensure that business processes fully leverage their robotics investments – production data flows up to ERP for analysis and costing, and business decisions flow down to automatically adjust robotic production. For example, if ERP detects a particular part is out of stock, it could put related robot jobs on hold via MES, preventing wasted effort.

In practice, achieving all these integrations can be challenging (as mentioned earlier, it ties back to the integration complexity). However, industry standards and architectures are making it easier. OPC UA (Unified Architecture), for example, is widely used as a communication standard to allow machines and software from different vendors to securely exchange data in a common format [32]. Many modern robot controllers now come with OPC UA servers, so an MES or IoT platform can readily pull data like cycle count, alarms, or current program status from any robot, regardless of brand, via OPC UA. On the higher level, standards like ISA-95 provide models for structuring data and workflows between enterprise, MES, and control levels, helping software engineers design integration points that align with best practices. The ultimate goal of such integration is a "single source of truth" and unified control: the product definitions (from PLM) drive the production activities (via MES to robots), and the results of production (from robots back through MES) inform business decisions (in ERP and beyond), all while digital twins and IoT analytics optimise the process continuously. In a fully integrated digital factory, one could trace a customer order through to the robot motions that built the product, and adjust any link in that chain with minimal manual intervention.

Industry Applications and Case Studies

Robotics and automation have been embraced across numerous industry sectors, each with its use cases and lessons learned. Here we highlight a few key sectors – automotive, electronics, pharmaceuticals, and logistics – to illustrate real-world applications of digital manufacturing with robotics:

Automotive Manufacturing

The automotive industry has been a pioneer in industrial robotics since the 1970s and remains one of the largest users of robots. In car assembly plants (the classic example of mass production), hundreds of articulated robots weld car bodies ("body-in-white" welding), apply paint coatings in paint shops, and perform tasks like sealing, dispensing adhesives, and assembly of heavy components. A modern car factory's welding line is often a mesmerising ballet of robot arms moving in concert to weld different chassis parts within a minute or two, with sparks flying – something only achievable with automation. Robots ensure high weld quality consistency and can work continuously in harsh welding environments (heat and fumes). Painting robots produce high-quality finishes while protecting human workers from toxic paint fumes. In engine and powertrain plants, robots handle hot castings, do precision machining, and even assemble engine components with high accuracy. Automation has allowed the automotive sector to achieve incredible productivity – for example, high-end factories can produce a finished car every 60-90 seconds. Automotive was historically so dominant in robot usage that until recently, it accounted for about one-third of all industrial robot installations globally[47]. (This share has been gradually declining as other sectors grow, but automotive is still a leader.) Beyond the factory floor, auto manufacturers are now using collaborative robots for tasks like assembling small sub-components or working alongside humans in final assembly (for tasks that are tricky to automate fully, such as threading a hard-to-reach bolt – a cobot can hold or pre-position parts to assist the human). A recent example from BMW involves deploying humanoid robots on assembly lines to add flexibility – these bipedal robots (in a trial phase) can perform various logistics and assembly tasks, illustrating the industry's continued innovation[42]. The automotive sector also heavily uses AGVs/AMRs in intralogistics: driverless systems deliver parts to the line just in time, and increasingly autonomous tuggers or carts follow dynamic routes in response to production needs. The concept of the "dark factory" (fully automated manufacturing) was first approached in automotive decades ago, though humans remain integral for many tasks in practice. Nonetheless, automotive showcases the full spectrum of robotics and automation working in unison, and has proven the benefits in throughput and quality (modern cars are produced with far fewer defects, thanks partly to robotic precision in manufacturing).

Electronics and Semiconductors

The electronics industry, including consumer electronics manufacturing and semiconductor fabrication, is another huge adopter of automation. In electronics assembly, precision and speed are critical – think of placing tiny electronic components onto printed circuit boards (PCB) or assembling smartphones, which have intricate parts. SCARA and delta robots are widely used here for ultra-fast pick-and-place of small components. For instance, high-speed delta robot mechanisms (commonly built into "chip shooter" machines) can place components at tens of thousands per hour with microscopic precision in PCB assembly. Robots also handle tasks like precisely dispensing tiny dots of solder paste or adhesive. In the final assembly of devices (like smartphones or computers), some companies have highly automated lines where robotic arms perform screwdriving, testing, and packaging. One striking example is Chinese electronics firm Xiaomi's "Dark Factory" for smartphone production. In this factory, robots and AI handle the entire assembly of smartphones without human operators, reportedly producing one phone every 30 seconds (or even every second for specific processes) with minimal human intervention[42]. This showcases how automation can achieve speed and quality – the term "dark" implies lights-out manufacturing, leveraging robotics to the fullest. Semiconductor manufacturing (chip fabs) also heavily rely on automation: wafer handling robots move silicon wafers between processing machines in ultra-clean environments (human intervention is minimised to avoid contamination). Automated material handling systems (overhead transport shuttles, sorters) move materials through the fab's many process steps. Even lab environments for electronics often use robot arms or gantry robots to handle delicate tasks under microscopes or in test equipment. The motivation in electronics is not just labour saving, but also precision (ensuring a high yield of good products when components are microscopic) and the speed needed to meet the enormous global demand for devices. As products miniaturise and become more complex (like smartphones with multiple cameras, sensors, etc.), robots are essential to assemble them reliably. Electronics manufacturing also pushes the envelope with collaborative robots in final assembly and AI vision systems for quality (detecting defects on tiny solder joints that humans can barely see). In summary, automation is deeply embedded in electronics production, from chip fabrication to device assembly, enabling the industry's high-tech products to be produced at scale.

Pharmaceuticals and Medical Devices

The pharma industry historically was not as automated as automotive, but this has changed significantly with advancements in robotics suitable for clean and sterile environments. In pharmaceutical manufacturing, maintaining sterile conditions and preventing contamination is paramount. Robotic systems are used for tasks like filling and packaging sterile vials and syringes, where isolator robotic arms can operate in a sealed sterile enclosure, doing the work without risking human contamination. This became especially important in vaccine production, where high throughput and sterility were needed (for example, during the COVID-19 vaccine manufacturing ramp-up, many facilities used fill-and-finish robotic lines). Robots also perform automated inspection of pharma products – e.g., machine vision systems check each vial for cracks or particulate, and robots reject any that fail criteria, much faster than human inspectors could[43]. Lab automation robots (sometimes specialised Cartesian robots or small articulated arms) prepare samples, pipette liquids, and manage assays, accelerating research and testing. Collaborative robots now appear in pharma labs to help technicians by handling repetitive pipetting or moving sample plates. Another area is pharmacy automation – robots in hospital pharmacies can pick and sort medications, prepare IV bags, etc., reducing errors and freeing pharmacists for clinical work. In medical device manufacturing, robots assemble devices like insulin pens, inhalers, or surgical instruments with high precision. For example, an assembly line for autoinjector pens might use several 6-axis robots to insert springs, syringes, and caps in a rapid sequence. The tolerance for error is low because these devices must be safe and reliable for patients; robots help ensure consistency. A cutting-edge case in biotech is cell therapy manufacturing: startups like Cellares have developed robotic cell culture systems (essentially a "cell therapy factory in a box") to automate the very labour-intensive process of producing personalised cell therapies [42]. This involves robots moving fluids, manipulating cell samples, and running analytic instruments in a sterile environment, aiming to make customised medicine scalable. The pharmaceutical and medical industries also benefit from digital twins and simulation, for instance, simulating a robotic dispensing system to ensure it meets FDA regulations before actual use. Overall, while people will always have a critical role in healthcare and pharma, robotics is making production processes safer (by reducing human handling of potent compounds), more efficient (24/7 production of drugs), and higher quality (fewer manual errors, with automation ensuring each dose is correct). The trend is so significant that even regulatory bodies have provided guidelines for using automation and data systems to ensure compliance (e.g., FDA's encouragement of automation to improve drug quality).

Logistics and Warehouse Automation

In the logistics, retail, and e-commerce sectors, robotics has seen explosive growth in the form of warehouse automation. With the rise of online shopping, fulfilment centres must process orders at breakneck speed, and mobile robots and automated sortation systems have become indispensable. A famous example is Amazon's warehouses, where the company has deployed over 750,000 mobile robots to assist in storing and retrieving products[48]. These robots (originating from Amazon's acquisition of Kiva Systems) carry shelves of products to human pickers, who then select items for orders – a drastic inversion of the old method where humans walked shelves. By automating travel, Amazon improved order pick rates and warehouse capacity. Today's warehouses use a variety of robots: AMRs with shelves or bins, robotic arms for picking individual items (using advanced vision and grippers to handle diverse products), automated conveyors and sorters that route packages, and palletising robots that stack boxes for shipping. Companies like DHL and FedEx use robotic arms to sort parcels or load trailers. Grocery distribution centres use robots to assemble pallets of mixed products for stores. In parcel logistics hubs, high-speed robotic systems can scan barcodes and divert packages faster than any manual process. Autonomous forklifts (self-driving forklifts or reach trucks) are starting to handle pallet moves in large warehouses, navigating safely alongside human-operated equipment. Moreover, drones or robotic delivery vehicles are being trialled for last-mile delivery in some cases. The benefits in logistics are speed and scalability: robots can handle surges in volume (like holiday seasons) by operating around the clock, and they reduce the physically strenuous work humans face (e.g., walking 15 miles a day in a large warehouse or lifting heavy packages repeatedly). Notably, human-robot collaboration is key here; for instance, in many fulfilment centres, humans still handle final item picking or packing because of the dexterity required, but robots bring the work to them (goods-to-person systems). The integration with inventory management systems (often part of ERP or dedicated WMS software) is critical – these systems direct robots and, in turn, update inventory counts as they move items. The logistics sector's embrace of automation shows how robots are not just for manufacturing products but also for moving products, effectively automating material flow and supply chain operations. As labour shortages hit warehouse jobs and as throughput demands increase, this sector is likely to continue leading in mobile robot and AI-based vision innovation. The sheer scale (hundreds of thousands of robots in use at single companies) demonstrates that automation has become a competitive necessity in logistics[48].

These examples barely scratch the surface – other industries seeing robotics growth include food and beverage (robots for picking and packaging food, slicing and sorting, etc., with hygiene considerations), aerospace (robots drilling and assembling aircraft components, which require high precision on large parts), construction (emerging robots for bricklaying or autonomous machinery), and agriculture (robotic harvesters, milking machines). Even in textiles and apparel, traditionally hard to automate due to floppy materials, robots combined with AI vision are starting to appear for tasks like fabric handling or sewing assistance. Each industry has its unique drivers, but the common theme is leveraging robotics and automation to improve efficiency, quality, and flexibility in response to market demands.

Importantly, case studies often show that success comes from tailoring the automation to the specific needs. For instance, an automotive plant may justify an expensive high-throughput robot line because volumes are huge. In contrast, a low-volume, high-customisation aerospace plant might opt for smaller collaborative robots that assist humans to get a mix of flexibility and automation. In pharmaceuticals, regulatory validation is crucial, so any robotic system must be thoroughly tested and replicated in digital form to prove reliability before going live. Lessons learned across industries indicate that a clear understanding of the process and a phased integration help in realising the gains of automation without disrupting production.

Current Trends and Future Outlook

The field of digital manufacturing is continuously evolving, and several key trends are shaping the next generation of robotics and automation solutions. These trends point toward systems that are more intelligent, more collaborative, and more seamlessly connected than ever:

AI-Driven Adaptive Automation: We are moving from robots that are programmed to those that can learn and adapt. Advances in AI, particularly machine learning, are enabling robots to handle variability and make decisions in ways they couldn't before. One aspect of this is adaptive automation – the ability of a system to adjust to changing conditions autonomously. For example, instead of stopping production when a part is slightly out of spec, an adaptive robot might adjust its grip or motion to compensate and still assemble the part successfully. Machine learning models can optimise robot paths for speed or energy efficiency beyond what human programmers might achieve. Reinforcement learning is being used for complex tasks like having a robot discover the best strategy to stack odd-shaped objects or to insert flexible parts. AI can also optimise scheduling and coordination: in a multi-robot assembly station, an AI controller might dynamically reallocate tasks between robots if one robot is performing more slowly due to some condition. Essentially, robots are increasingly aware of their performance and environment and can self-optimise. We see this with some modern cobots that auto-tune their force controls by feeling their way through an assembly process, or with AI vision systems that improve their defect detection as they see more examples (continuously learning on the line). The promise is that automation will handle high mix and variation more gracefully, which historically has been a weakness. One concrete example: force-sensitive assembly robots that can adapt to slight misalignments – AI algorithms allow them to feel when a peg isn't going in a hole and wiggle it until it fits, much like a human would. Another example is predictive optimisation: if an AI notices a specific pattern (say, every day around noon, a process slows down), it might deduce a cause and adjust parameters before it happens. A recent trend is also using Generative AI to assist in automation software development – e.g., generating robot programs or PLC code from high-level descriptions, making automation engineering faster—all these point to more autonomous and resilient production. As one industry piece puts it, manufacturers are seeking "flexible, productive, and adaptive automation" more than ever to cope with rapid market changes[(49)]. We can expect future production lines to rebalance themselves in real time automatically. If one station has a problem, the system finds an alternate path or way to fulfil the schedule, with minimal human intervention.

Human-Robot Collaboration (HRC) and Cobots: The trend of human-friendly robots is growing. Collaborative robot arms are becoming more capable (higher payloads, faster speeds with safety), and new forms of collaboration are being explored. Beyond cobot arms, we have mobile robots that work in human spaces (warehouse robots rolling among human pickers, or cleaning robots in facilities) and even exoskeletons or wearable robots that assist human workers. The idea is to create a hybrid workforce where robots and humans each do what they're best at. Humans excel at creative thinking, complex decision-making, and fine dexterous tasks; robots excel at endurance, precision, and strength. In a collaborative setup, you design the workflow such that neither is overburdened. For instance, a human might perform the tricky assembly of cables in a product while a cobot holds the product and then does the screwing operation – each partner contributes. Collaboration also means easier interaction interfaces – workers can guide a robot arm to teach it, or use voice or AR (augmented reality) instructions to command robots on the fly. The future likely holds more fenceless manufacturing cells where humans and robots freely share space. This provides flexibility like a manual line (where people can walk up and do something), combined with the efficiency of automation. With sensor and vision technology improving, robots can have a constant awareness of humans nearby and adjust accordingly – e.g., slow down if someone comes close, or coordinate moves to avoid each other. Standards and safety certifications for cobots are evolving to support this growth (ISO 10218 and ISO/TS 15066 define safety requirements for collaborative robotics). Industry adoption is climbing; collaborative robots made up about 10.5% of new industrial robot installations in 2023 (a figure that has grown year over year)[41]. This trend is expected to continue as cobots open automation to new types of operations and smaller companies. We will also see AI in cobots, enabling them to understand intent (perhaps via gesture recognition or advanced vision) and become more intuitive co-workers. An example of emerging tech is robots that can safely pass objects to humans or vice versa, which is helpful in assembly lines or even healthcare (a nurse robot passing tools to a surgeon). Human-robot collaboration extends to planning as well – e.g., systems where human supervisors set goals and robots use AI to figure out the details. The future factory might look less like rows of isolated robots and more like a fluid space of humans and robots constantly negotiating tasks in real time.

Cloud Robotics and IIoT Connectivity: As mentioned earlier, connectivity is a big theme. Cloud robotics refers to robots leveraging cloud infrastructure for data and computation. Instead of each robot working in isolation, they share a brain (or memory) in the cloud. This means, for example, that if one robot in one factory learns something, another robot elsewhere can immediately benefit. Imagine an AI vision model identifying a new defect – once updated in the cloud, all connected robots globally could now catch that defect. Cloud computing can also offload heavy computations from the robot, such as path planning, vision processing, or AI model inference, to powerful servers, sending results back to the robot in milliseconds. This can make robots more powerful without needing expensive on-board computers. Cloud connectivity also simplifies centralised monitoring: a company with multiple plants can monitor all its robots' status and performance through a cloud dashboard, perform over-the-air updates to their programming, and even control them remotely if needed. Some vendors offer cloud-based robot management platforms where you can simulate, program, and deploy to robots from a web interface. The collective learning aspect is fascinating: fleets of robots (like warehouse AMRs) can share data about their environment to improve navigation routes or avoid hazards continuously. A concrete example is a fleet of AMRs in a warehouse. If one encounters a temporary obstacle (like a split load), it can inform the fleet, and the cloud route optimiser will redirect others until it's cleared. Cloud robotics also aligns with the IIoT push for every device to be networked. There are challenges, notably latency and security, but advancements like 5G networks (with low latency) and robust encryption are helping. As one reference notes, cloud robotics allows robots to offload tasks like path planning and object recognition to cloud servers, enabling more complex tasks to be performed by simple robots[50]. A future vision is a kind of "robot app store" – robots could download skills on-demand from the cloud (for instance, a new AI skill to identify a different part). Some industrial players and consortia are actively working on frameworks for this. We might also see edge-cloud hybrid architectures, where some processing is near the robot (for immediate response) and more complex learning and analytics are in the cloud. This trend will likely accelerate with the proliferation of 5G/6G in factories, as wireless high-speed connectivity becomes standard, untethering robots from fixed control points.

Advanced Sensing and Vision, incl. 3D Vision: A related trend is the improvement in robot perception. Cameras, lidar, radar, force sensors, and even new tactile sensors are giving robots richer information about their environment. 3D vision allows robots to understand the shape and depth of objects, which is crucial for tasks like bin picking, where items are randomly piled. Recent strides in deep learning mean vision systems can recognise a huge variety of objects and even estimate their weight or centre of gravity to help a robot pick them correctly. Sensor fusion (combining inputs from vision, force, audio, etc.) will make robots more situationally aware. For instance, a robot could "hear" acoustic feedback if a part is seated correctly (by sound frequency) or use laser scanners to dynamically map its surroundings for navigation (in the case of mobile robots). These advanced sensors feed into AI models (as discussed above) to drive more autonomous behaviour. Another aspect is better feedback for humans – AR/VR tools can let human operators see what the robot "sees" or intends, improving trust and ease of programming. Overall, more advanced sensing is an enabler for many of the other trends (adaptation, collaboration, etc.), as it's the foundation of understanding the world beyond pre-programmed knowledge.

Standardisation and Interoperability (OPC UA, ISA-95, ROS Industrial): As automation systems become more complex and multi-vendor, the need for standards to ensure seamless integration is critical. We've already mentioned OPC UA, which has emerged as a key standard for machine data interoperability in Industry 4.0. OPC UA provides a universal language for industrial devices to communicate; many robots now support it so they can easily integrate with different PLCs or MES without custom drivers[51]. There are even OPC UA companion specifications specifically for robotics, defining standard information models for robot statuses, alarms, etc., to simplify integration. ISA-95 (and the related IEC 62264) continues to guide how enterprises structure their automation hierarchy and data exchanges, ensuring clarity between levels (ERP, MES, control) and enabling modular integration of systems. Following ISA-95 can future-proof a deployment because it aligns with how most off-the-shelf software expects to interface. Another standard to mention is ISA-88 for batch process control, relevant in process industries, which is also often used in MES design; robots in batch processes (like pharma) might be integrated following ISA-88 models of recipes and procedures. On the robotics software side, ROS (Robot Operating System) Industrial is an initiative adapting the open-source ROS framework for industrial use. This is enabling more standardised development of robot applications across different robot brands. In the future, one might more easily deploy the same code to a FANUC or an ABB robot if they both support a common ROS-I interface, for example. Standardisation also touches on safety, ensuring robots from different vendors all meet certain safety performance levels so they can work together without gaps. For mobile robots in warehouses, emerging standards like VDA 5050 (in Europe) define how fleet management systems can control AMRs from multiple manufacturers in one system. All these standards and protocols aim to avoid vendor lock-in and reduce integration effort, which will be increasingly important as factories might have dozens or hundreds of robots of various types needing coordination. In essence, the future is moving towards open, interoperable "plug-and-play" automation, where deploying a new robot or device is not a bespoke IT project but a quick configuration because all the interfaces are standardised. This will significantly reduce the engineering cost and complexity currently associated with automation, thereby accelerating adoption.

Scalability and Modular Production: Future factories are likely to be modular, with production lines that can be reconfigured like building blocks. This goes along with flexible automation: instead of huge monolithic lines, we might have modular cells with robots that can be rearranged or repurposed as needed (sometimes referred to as "Lego" factories). Standardised docking interfaces for mobile robots or cobot workstations mean you can add capacity by literally adding another module. Scalability also refers to taking pilot projects and rolling them out across multiple sites – a big focus now is on scaling automation from one prototype cell to an entire enterprise. This is where digital twins and simulation help (you can model a scaled-up system and ensure it will work before replicating 10 cells). Additionally, cloud management helps scale – if you have 1000 robots, you need centralised oversight to update and manage them efficiently, which is why robotics fleet management software is a hot area. The modular approach aligns with the trend of mass customisation – a modular, flexible line can switch between many products quickly, enabling economical production of highly customised goods.

Looking forward 5-10 years, we can envision "smart factories" that are largely self-organising. Raw materials might arrive with RFID tags telling the factory what final product they need to become; AI-driven scheduling systems allocate tasks to various robotic cells; mobile robots deliver the parts to the right cell at the right time; collaborative robots and humans work on assembly tasks together for maximum efficiency; quality is monitored in real-time by vision systems and any deviation triggers immediate adjustment; and throughout this process, all data is synchronized to digital models and the cloud for continuous optimization. Such factories will be highly resilient, able to continue production despite disruptions by re-routing tasks (a concept sometimes called "biological manufacturing", where the system adapts like an organism). Edge computing with AI might handle instant control, while cloud computing does heavy analytics in the background. We will also see more domain-specific AI – for instance, AI specifically trained on decades of welding data to guide robotic welding better than any human welder could. And in terms of hardware, robots may become more varied in form: multi-armed robots, soft robots for gentle material handling, micro-robots for very small-scale assembly, etc., depending on needs.

For manufacturing professionals, staying abreast of these trends is crucial. The competitive advantage often lies in how effectively an organization can incorporate new technologies to improve its operations. Companies that successfully blend human expertise with cutting-edge automation – all under a unified digital strategy – will likely lead in productivity and innovation. The future outlook for robotics in digital manufacturing is undeniably exciting, with a convergence of technologies creating smarter and more capable production systems than ever before.

Key Robotics Vendors and Platforms

The industrial robotics landscape is populated by several key vendors that offer a range of robot models and automation solutions. Each has its specialties and ecosystem of software and support. Here we outline some of the major players and platforms that manufacturing professionals should be aware of:

ABB: ABB is a Swiss-Swedish multinational and one of the "big four" industrial robot suppliers globally. They produce a wide variety of articulated robots (from small 6-kg robots up to heavy 800+ kg payload robots) used widely in automotive and general industry. ABB robots are known for their reliability and advanced motion control. They also have a strong presence in paint robots and have offerings in SCARA and delta robots (e.g., ABB's FlexPicker delta robot for fast picking). ABB was a pioneer in collaborative robots with the introduction of YuMi, a dual-arm small cobot designed for electronics assembly. ABB's robotics platform includes the IRC5 robot controller and programming language RAPID, as well as simulation and offline programming software RobotStudio, which is popular for digitally testing robot programs. ABB has also expanded into mobile robotics by acquiring ASTI Mobile Robotics, indicating a push towards integrated solutions that combine stationary and mobile automation. ABB's strength lies in providing complete digital solutions – they integrate robots with PLCs (they have their own PLC line), machine vision, and even MES systems (via their manufacturing operations suite). Their products are common in automotive plants, but also in 3C (computer, communications, consumer electronics) manufacturing, food and beverage (palletizing robots), and more.

KUKA: KUKA is a German robotics company known for its bright orange robot arms. They have historically been very strong in automotive welding and assembly – many car factories in Europe and Asia use KUKA robots extensively. KUKA offers a full range of 6-axis robots, including some of the largest payload robots in the market (suitable for tasks like lifting car bodies or positioning heavy fixtures). They also have SCARA robots (via a subsidiary) and a collaborative robot line called LBR iiwa (a high-end force-controlled cobot that was one of the first in its class). KUKA is notable for some innovative applications – for example, they have robots used in entertainment (like robot arms moving people for rides) and in medical (radiation therapy machines). Their robot control software is KRL (KUKA Robot Language) running on the KRC controller, and they offer the KUKA Sim software for simulation. One differentiator is KUKA's push towards mobile manipulation: they've demonstrated robots on autonomous platforms (the KMR iiwa – a mobile base with a cobot arm) for flexible factory logistics. KUKA was acquired by the Chinese firm Midea in 2016, which has led to some increased presence in China's automation market. KUKA also provides turnkey automation solutions through its systems engineering division. They emphasize the concept of the "robotic cell of the future" where robots can be easily reconfigured, something they call Industrie 4.0 readiness.

FANUC: FANUC is a Japanese company and by some measures the largest industrial robot manufacturer in the world (in units deployed). FANUC robots are iconic in their yellow color and are known for their exceptional reliability and longevity – there are reports of FANUC robots running for decades with minimal issues. They have a vast product line: from tiny delta and SCARA robots for picking small parts, up to huge gantry robots for heavy material handling. FANUC is extremely dominant in the automotive sector (especially in Asia and North America) and is also a major supplier to electronics manufacturing. They have a series of collaborative robots (the FANUC CR and CRX series, green-colored to differentiate, designed to stop safely on contact). One of FANUC's strengths is in machine tool automation – since FANUC also makes CNC controllers, their robots integrate well with CNC machines for loading/unloading parts (their Robot Drill line is an example of combined machine tool + robot). FANUC's software ecosystem includes the ROBOSHOT (injection molding integration), ROBOGUIDE simulation, and ZDT (Zero Down Time), a cloud-based platform for predictive maintenance of robots. They have embraced IIoT; for example, ZDT monitors thousands of robots in the field to predict failures[45]. FANUC's controllers use the Karel programming language and TP (Teach Pendant) programming, which is widely taught. As a highly conservative but dependable brand, FANUC emphasizes standardization and high-volume manufacturing; their factories in Japan are famously almost fully automated, with robots making more robots.

Yaskawa Motoman: Yaskawa (brand Motoman in many markets) is another Japanese giant of robotics. Yaskawa is known for a broad range as well, historically strong in spot welding robots for automotive and in general material handling robots. They have a presence in industries like packaging, and they were also among the first to explore dual-arm robots (the Motoman SDA dual-arm series). Yaskawa's collaborative robot line is the HC series. Their platform includes the DX200, YRC1000 robot controllers and programming via Inform language. Yaskawa has a reputation for cost-effectiveness and strong performance in welding and handling. Many system integrators use Yaskawa robots as a base for palletising or welding cells. Yaskawa also produces drives and servo motors, and they leverage that motion expertise in their robots (it's a common thread that many robot companies either came from the machine tool world or the motion control world). In recent trends, Yaskawa has been working on human-assistive robots and easy-to-program interfaces to target SMEs.

Universal Robots (UR): UR, a Danish company, is a pioneer in collaborative tabletop robots. Founded in 2005, UR essentially created the market for smaller cobots that companies could use without a robotics expert. Their UR3, UR5, UR10 (and later UR16, etc., numbers roughly indicating payload in kg) are widely used in industries like electronics, small parts assembly, machine tending, and even services (like barista robots or medical lab automation). An extremely user-friendly interface characterises UR cobots – they can be taught by demonstration and via a simple GUI on a tablet, in addition to traditional programming. They intentionally have lower speed and payload, which makes them intrinsically safer around people (they are limited-force as well). One key aspect of UR's success is its ecosystem, UR+, where third-party developers offer plug-and-play grippers, vision cameras, and software plugins that integrate seamlessly with UR robots. This ecosystem approach has made it easy to tailor UR cobots to specific tasks (for example, a plug-in for interfacing a UR robot with a particular brand of CNC machine, or a ready-made kit for palletising with a UR). Many small and mid-sized enterprises have adopted UR as their first foray into robotics, due to this approachability. While UR robots are not as fast as traditional ones, their value is in flexibility and ease of use. They have seen deployment in automotive as well for supplemental tasks, and even in healthcare (some UR arms were used to automate COVID-19 testing labs, for instance). UR has inspired dozens of other companies to launch cobots, but they remain a market leader in that segment.

Omron (Techman, Adept): Omron is a Japanese automation company (famous for sensors and PLCs) that has also entered robotics through acquisitions. They acquired Adept Technology (an American robotics firm known for SCARA robots and mobile robots) and Techman Robot (a Taiwanese cobot maker). As a result, Omron's portfolio now includes Adept's SCARA robots (Omron Adept SCARAs are common in electronics and food packaging), Autonomous Mobile Robots (AMRs) initially from Adept's Lynx platform (now Omron LD series, used for internal transport), and collaborative robots (Techman's cobots, which interestingly have an integrated vision system in the robot arm). Omron's strategy is to provide a total automation solution from sensor to robot to control to safety – they can supply an entire cell's automation components. For example, Omron might provide the safety light curtains, the PLC, the vision system, and the robot all together. Their mobile robots are notable; Omron (Adept) was one of the first to do intelligent AMRs for factories, and these are used in semiconductor fabs and other intra-factory logistics. Omron Techman cobots come with built-in vision on the robot's wrist, making hand-eye coordination more straightforward out of the box. As Omron is strong in machine vision, they integrate that with their robots for tasks such as inspection. Omron's PLCs (the NJ/NX series) even have robotics functions to synchronise motion between conveyors and robots tightly (essential for pick-and-place on fast-moving lines). So, Omron is an example of a vendor bridging fixed and mobile robotics with the larger automation environment.

There are several other important companies:

- Mitsubishi Electric: Offers industrial robots (especially SCARAs and small articulated robots) often tightly integrated with their PLC and CNC offerings. Strong in electronics and pick-and-place.

- Epson Robots: A leader in SCARA robots with very high precision, widely used in electronics assembly and slight part handling.

- Staubli: A Swiss company known for high-precision robots, including 4-axis and 6-axis robots that are designed for cleanroom or harsh environments (their robots are common in pharma and solar panel manufacturing, and even in environments with strict hygiene like food, as they have washdown-ready designs). Staubli also makes unique coupling devices and is known for very reliable engineering.

- Kawasaki Robotics: Japanese company (not the motorcycle division, though related historically) that offers a full range of industrial robots, known for heavy payload and automotive applications as well as a presence in life sciences (they had a medical robot partnership). They have been in robotics for a long time and have many automotive clients.

- Nachi-Fujikoshi: Another Japanese firm producing robots as part of their machining business. They have a respectable install base in Asia.

- Hanwha, Doosan (Korea): In the collaborative robot boom, some new entrants like Doosan Robotics have released cobots with high payloads and advanced features, gaining some market share, particularly in Asia.

- Rethink Robotics (legacy): Known for the Baxter and Sawyer cobots that were iconic in the early 2010s for their friendly design. However, the company closed in 2018; its technology and brand were acquired and may re-enter via different channels.

- Mobile robot specialists: Aside from Omron, companies like MiR (Mobile Industrial Robots, a Danish firm now under Teradyne, which also owns UR) are leaders in AMRs; their robots are often used in flexible manufacturing logistics. Amazon Robotics (for warehouse bots) is mostly in-house to Amazon, but their innovations lead the way for that sector. Chinese firms like Geek+ and Quicktron are becoming notable in AMRs and warehouse robots globally.

- System Integrators and Solution Providers: It's worth noting that beyond robot manufacturers, there is a vast ecosystem of system integration companies (like JR Automation, FFT, etc.) that create turnkey solutions using these robots, and they often develop their platform of expertise (for example, an integrator might specialize in robotic painting systems with custom software, using standard robot arms as a base). These integrators effectively act as vendors when you are shopping for a complete automated line.

Each vendor often has its programming environment and quirks, so one consideration for manufacturers is whether to standardise on one brand for ease of maintenance or to mix and match to use each robot where it's best. Increasingly, cross-vendor compatibility via standards (or middleware like ROS Industrial) is reducing the friction. Also, many vendors partner – e.g., you might see a FANUC robot with a Cognex vision system and a Rockwell (Allen-Bradley) PLC controlling it; these companies ensure their products can work together for such common combinations.

Another concept is robotics platforms/software beyond hardware. For instance, Siemens (while not a robot maker) offers the Siemens OpCenter (formerly SIMATIC IT) MES and TIA Portal/PLC systems that tie in robots via PLC function blocks and OPC UA, effectively offering a platform to integrate any robot brand. Rockwell Automation similarly provides an integration environment (and through alliances can make any robot behave in a unified way in their Logix PLC platform). There are also startups focusing on "universal" robot programming platforms and simulation tools that can program multiple brands.

Finally, mention should be made of the International Federation of Robotics (IFR), which, while not a vendor, is the industry body that tracks robotics statistics and trends. They release the annual World Robotics report, which often cites the top industries and countries for robot use, and sometimes references market share of vendors or technology trends. According to IFR data, the top five countries for industrial robots (by installation) are typically China, Japan, USA, South Korea, and Germany – reflecting both manufacturing volume and technology adoption rates.[38]. Such data gives context that the named vendors are often multinational: e.g., ABB is strong in Europe and the Americas, FANUC and Yaskawa in Asia, etc., but all are global now.

For a manufacturing professional, familiarity with at least one or two of the major robot brands and their programming is helpful. Many educational programs teach on a particular brand, but concepts usually carry over. Knowing the ecosystem (such as which vendor has good local support or integrator availability for your needs) is also essential when choosing a partner for automation. Equally, one should consider the software and controller capabilities – e.g., if you need vision-guided motion, ensure the vendor easily supports that; if you plan to simulate offline heavily, check that the vendor's simulation tools are robust.

In summary, the robotics vendor landscape offers a rich selection, with legacy giants providing proven, scalable solutions and newer entrants or focused specialists offering innovative or niche capabilities. The good news is that competition and collaboration among these vendors drive continuous improvement in technology and more options for end-users to find the right fit for their digital manufacturing needs.

Best Practices for Planning, Deploying, and Scaling Automation

Successfully integrating robotics and automation into a factory requires more than just buying robots – it demands careful planning, skilled deployment, and thoughtful scaling. Many projects have faltered due to underestimating the organisational and technical work needed. Here we outline best practices and guidelines that experienced manufacturers and integrators recommend to ensure that automation yields the desired benefits:

Define Clear Objectives and Identify Suitable Tasks: Start with why you want to automate. Is it to increase the throughput of a bottleneck process? Improve the quality of a step prone to human error? Reduce labour in a hazardous area? Clear objectives will guide your project and metrics for success. Perform a thorough analysis of your current processes to identify the best candidates for automation[38]. Good candidates are typically tasks that are repetitive, standardised, high-volume, and do not require complex judgment. Also, consider tasks causing bottlenecks or safety issues. Involve process engineers and line workers in this identification; they often know pain points intimately. It's usually wise to start with a pilot on a task that has high impact but manageable scope, rather than tackling the most complex process first. Evaluate feasibility: for each candidate task, ask if a robot can do this better, faster or safer? [38]. For example, a robot might excel at a welding job with consistent parts. Still, if your parts vary too much or the task needs delicate custom fitting, it may be very hard to automate without first changing the product design or process. Avoid automating just for the sake of it – ensure the task is appropriate. Sometimes, a simple poka-yoke or ergonomic fix for a human process is better than full automation if the task is rare or highly variable.

Involve Stakeholders and Get Expertise Early: Automation projects should involve cross-functional stakeholders from the outset – production managers, engineers, operators, maintenance, IT, and even finance (for ROI analysis). Early buy-in from those who will use and maintain the system is crucial. If the in-house team lacks experience with robotics, engage with external system integrators or consultants early on. They can provide input on solution concepts and realistic budgets. Many companies partner with a reputable robotics integrator who designs, builds, and commissions the robotic cell. It's essential to check the integrator's track record in similar applications. When evaluating vendors (robot manufacturers or integrators), consider not just price but compatibility with your existing systems and standards[38]. Does their solution integrate with your MES? Do they use open standards or proprietary systems? Can their equipment work with your existing sensors or conveyors? Favour solutions that adhere to common standards (Ethernet/IP, OPC UA, etc.) to ease integration. Additionally, verify that any vendor selected offers good training and support – you are entering a long-term partnership, so their responsiveness and ability to support you post-installation matter immensely[38]. Do not skip the phase of detailed proposal and simulation – good integrators will often simulate the robotic cell to prove it can meet cycle time and fit in the space. Insist on seeing such evidence or even visiting a reference installation.

Plan the Cell Layout, Safety, and Workflow Thoroughly: The design of the robotic workcell or system is a critical engineering step. Ensure the layout has proper safety measures – including fencing or sensors for traditional robots, or safety-rated sensors for cobots if needed. Conduct a risk assessment early: identify all potential hazards (pinch points, reach of the robot, failure modes) and mitigate them (safeguards, emergency stops, etc.)[52]. Many countries require compliance with ISO/ANSI/RIA safety standards – make sure your design meets these. Plan how materials will come in and out of the cell (part presentation is often a challenge – you may need bowl feeders, conveyance, etc., to get parts to the robot in a repeatable way). Consider the end-of-arm tooling (EOAT) design carefully: the gripper or tool the robot uses must be reliable and robust, as it directly affects the success of operations. Also, design for maintainability: can an operator easily replace a worn gripper jaw or clear a jam? For cobot implementations, even though they are inherently safer, still train personnel and mark zones if needed; complacency around cobots can still lead to accidents if the tool or part is sharp, for example. Another key aspect is the workflow: map out each step in the automated cycle, including any handoffs between humans and robots. Ensure there are no ambiguous responsibilities – e.g., if a robot places a part and a human tightens it, make sure sensors or poka-yokes confirm that before the robot moves on. Use an MES or at least a simple scheduler to coordinate if multiple pieces of equipment interact. Essentially, think through the entire operation from raw part to finished part, not just the robot motion in isolation. It's often valuable to run a simulation or even a physical dry run with mock-ups (like cardboard layouts) to verify the concept.

Pilot the System and Train the Team: Before full production rollout, test the automated system thoroughly. This may involve a pilot running in parallel to the manual process until it's proven. Use this pilot phase to allow operators and maintenance staff to get familiar with the system and provide feedback. Training is crucial: operators should know how to start/stop the system, recover from simple errors (like resetting the robot if it faults out, clearing any obstructions, etc.), and the basic HMI (human-machine interface) interactions. The integrator or vendor should train maintenance technicians on preventive maintenance tasks (lubrication, replacing parts, etc.) and basic troubleshooting (diagnosing error codes, sensor alignment, etc.). Some companies choose a few tech-savvy operators to get advanced training in robot programming so they can make minor tweaks or touch-ups to the program when needed. Keep training materials (and backups of all programs/configurations) on-site and accessible. A best practice is to create standard operating procedures (SOPs) for the automated cell, including step-by-step procedures for startup, shutdown, changeover to a new product (if applicable), and emergency recovery, and train everyone on those. It's also wise to prepare for the "what-ifs": have a contingency plan if the robot is down – can you revert to manual process in the short term? During the pilot, gather data: cycle times, downtime reasons, etc., and compare to your goals. Often, adjustments are needed – maybe the robot program needs optimisation to reach the target cycle time, or an extra sensor is required to detect a rare fault condition. It's normal to iterate.

Gradual Ramp-Up and Change Management: When deploying the automation live, do a gradual ramp-up. Perhaps run it during one shift while keeping the manual process as a backup in another shift, until confidence is built. Monitor performance closely. Celebrate quick wins (e.g., quality improved, or throughput met) to show the team the positive impact. Manage the change with the workforce: clear communication that robots are tools to assist, not replace, indiscriminately, will help reduce fear. If some job roles will change (e.g., assemblers becoming cell monitors or technicians), work with HR to possibly reskill those whose roles are affected (maybe even involve them in the automation project, so they feel ownership rather than displacement). It's often noted that engaging the workforce turns them into champions of the new technology rather than adversaries. Give operators a channel to provide suggestions – since they usually know the process deeply, their insights can improve the automated solution or catch issues early.

Leverage Digital Tools (Simulation, MES, Monitoring): Use digital manufacturing tools to your advantage. Simulation and offline programming were mentioned – they can save time and prevent costly mistakes by debugging the system virtually. Digital twin models can be maintained even after deployment – for example, simulate any proposed changes in the robot's operation on the digital twin to assess the effects. If using MES or other software, integrate the cell to receive real-time data and track metrics such as OEE (Overall Equipment Effectiveness). Real data will help you quantify the improvement and also catch issues (if OEE is lower than expected, you can drill down into the causes). Consider using monitoring and alerting systems: e.g., if a robot triggers a fault and stops, have it notify maintenance via a text or alarm. Many modern robots allow remote monitoring – ensure it's set up (with IT security in mind). Data from the system can also highlight bottlenecks that shift elsewhere once this task is automated; be ready to address the next bottleneck (automation is often a continual journey of finding the following weakest link).

Optimise and Standardise: Once the robotic cell is stable, look for optimisations. This could be fine-tuning robot motion to shave off a second here or there (sometimes using the robot's own analytics or AI add-ons to optimise the path), or improving how parts are fed to reduce any slight delays. Over the long run, minor improvements yield significant gains. It's also wise to document the lessons learned from this deployment. Create internal standards if you plan to do more automation: for instance, standardize on certain brands or safety systems if they worked well, develop a preferred architecture (like "our robots will talk to our PLC via OPC UA using this data model"), and maybe even mechanical design standards (such as all new fixtures should allow robot access from a certain angle etc.). This will make future projects faster and more consistent. If you have multiple plants, share the knowledge and possibly replicate successful cells elsewhere, but also be mindful that not all sites are identical (each might need some localisation or adjustments).

Scaling Up Deployment: After a successful pilot, plan the roadmap for scaling automation in your factory or company. Prioritise projects by ROI and strategic value. It can be beneficial to maintain an automation roadmap that aligns with your production forecasts and product plans. For scaling, ensure your infrastructure can handle it: for example, do you have the electrical capacity for several more robotic cells? Is your network robust for many more connected devices? Work closely with IT/OT departments to ensure scalability of the digital infrastructure (for instance, using proper network segmentation for all the new devices to avoid bandwidth or security issues). From an organisational standpoint, as you scale, you might need to establish a dedicated automation engineering team or upskill more technicians to reduce reliance on external integrators. Some companies create a "Centre of Excellence" for robotics that supports all factories. Also, budget for spares – as you have more robots, having critical spare parts on hand (like a spare robot controller or extra servo motors) can drastically reduce potential downtime across the fleet.

Maintenance and Continuous Improvement: Treat robots as high-value assets that need care. Implement a preventive maintenance schedule in your CMMS (Computerised Maintenance Management System) for each automation cell – including lubrication, bolt tightening, calibration checks, battery replacements (many robots have encoder backup batteries), etc., as recommended by the manufacturer. Also, maintain the peripheral devices (sensors, conveyors). Monitor key indicators like a robot's motor torque trends or temperature – abnormal trends might indicate issues such as grease deterioration or wear. Many robots come with diagnostics – use them (or an external solution like vibration analysis) to avoid unexpected failures. Keep the robot programs and parameters backed up offline in case of a hardware failure (this is often overlooked until a robot controller fails, and it's discovered that the only program copy was on the controller). Also have a recovery plan if something goes wrong – e.g., if the robot crashes or an alignment is off, have a documented method to re-teach or recalibrate. Continuous improvement should be a mindset: gather operator feedback, check quality data pre- and post-automation to ensure the automated process indeed maintains or improves it. If new product variants are introduced, plan how to update the automation (maybe invest in quick-change tooling or flexible fixturing). There should be a feedback loop where any issues that occur (e.g., a particular sensor frequently gets dirty and stops the cell) are analysed and a solution implemented (maybe add an air nozzle to blow it clean periodically, or improve the sensor placement). Over time, your team will become more proficient and proactive, and the automation will run more smoothly and efficiently.

Consider Scalability to Future Technologies: Plan with the future in mind. For example, if you might want to add AI vision later, maybe choose a robot and controller that can integrate that later, or leave space in the cell for additional cameras. Or if you anticipate possibly moving to a new facility, design modular cells that can be relocated. Keep an eye on evolving standards – if something like a new safety scanner or communication protocol could simplify your setup, be ready to adopt in future expansions. Scalability is not just more of the same, but also the ability to upgrade. This is where using standardised, open platforms pays off, as you can integrate new tech more easily. For instance, if all your robots feed data into a unified database, when you deploy a fancy new analytics AI, it can train on data from all cells consistently.

In summary, the best practices revolve around good project management (plan, do, check, act cycle), people enablement (training and buy-in), and technical diligence (safety, integration, maintenance). A well-executed automation project can significantly boost a factory's performance, but a poorly executed one can become a costly headache. Following these guidelines helps stack the odds in your favour for a smooth transition to a more automated, digital manufacturing operation. Remember that automation is not a one-time project but a continuous journey – start small, learn, and build momentum. By adhering to these best practices, manufacturers can more reliably harness the power of robotics and automation to create efficient, flexible, and competitive digital factories. The payoff is not just in immediate cost or output metrics, but in building a resilient operation ready to adapt to the future of manufacturing.[38] [49]

Cyber-Physical Systems

Cyber-Physical Systems (CPS) are engineered integrations of computation and physical processes, tightly interconnected through networking and control. In essence, a CPS merges the cyber (computing, software, data) with the physical (machines, devices, sensors/actuators) into a unified system. The U.S. National Science Foundation defines CPS as engineered systems built from and dependent upon the seamless integration of computational and physical components[53]. Similarly, NIST describes CPS as "smart systems that include engineered interacting networks of physical and computational components", noting that such highly integrated systems enable new functionalities across domains like smart manufacturing[54]. In a CPS, the physical equipment (e.g. robots, conveyors, machine tools) is equipped with embedded software and sensors, connected via networks to computing infrastructure that monitors and controls the physical processes in real time.

Key Characteristics of CPS

According to Germany's acatech (National Academy of Science and Engineering), cyber-physical systems exhibit several distinguishing features[54]:

Sensor-Actuator Integration: CPS directly sense physical conditions through sensors and affects physical processes through actuators[54]. For example, a CPS in a factory might measure temperature, vibration or product dimensions and adjust machine parameters on the fly.

Data Processing and Control: They continuously evaluate and save data, and interact with both the physical environment and digital networks, either autonomously or with human input[54]. This means the system can make real-time decisions (e.g. adjusting a robot's speed) based on sensor data.

Network Connectivity: CPS components are typically interconnected via networks, often Internet-enabled, allowing them to communicate with each other and with higher-level information systems[54]. Each element (machine, device, etc.) in a manufacturing CPS can have its IP address and communicate status or receive commands.

Use of Global Data/Services: They can leverage externally available data and cloud services, integrating, for instance, supply chain information or predictive analytics from remote servers[54].

Human Interaction: CPS include human-machine interfaces to allow operators and engineers to monitor and influence the system[54]. In a plant setting, this could be dashboards showing machine health or control stations for manual overrides.

These characteristics illustrate that a CPS is more than just an automated machine – it is a networked, intelligent system that blurs the line between the digital and physical realms. The concept of CPS was first popularised in the mid-2000s (the term is often credited to Dr. Helen Gill of NSF[54]) and has since become foundational in discussions of modern smart manufacturing and Industry 4.0. CPS technology allows physical processes to be not only automated but also to be represented in the digital domain (sometimes as "digital twins"), where they can be analysed, optimised, and even controlled remotely in real-time[54, 55].

Cyber-Physical Systems in manufacturing might consist of machines embedded with sensors and microcontrollers, all connected through an industrial network to manufacturing execution systems or cloud platforms. Through this integration, the CPS can monitor itself, make decisions or adjustments, and coordinate with other systems – essentially creating a smart factory environment where physical production is tightly coupled with digital computation and data exchange.

CPS Integration in Digital Manufacturing

Digital manufacturing refers to the use of digital tools, data, and connectivity throughout the production lifecycle – from design and simulation to production and supply chain management. CPS plays a pivotal role in enabling this digital transformation on the factory floor. By embedding connectivity and intelligence into machinery and production lines, CPS serve as the "nervous system" of a digital factory, linking the physical operations with computational decision-making.

In practical terms, integrating CPS into manufacturing means that formerly isolated machines or process steps become part of a connected network. For instance, consider a smart production line where each robot and conveyor is CPS-enabled: sensors on these machines feed operational data to a central platform, which in turn sends control commands or adjustments back to optimise the process. This real-time feedback loop is fundamental to digital manufacturing. It allows production systems to self-adjust and coordinate without always needing human intervention. One industry perspective defines a manufacturing CPS as an "internet-enabled physical entity, such as a pump or compressor, embedded with computers and control components (sensors and actuators), capable of self-monitoring and communication"[54]. In other words, every significant piece of equipment can report on its status and collaborate with others through networking.

Industry 4.0 and IIoT Context

CPS integration is often discussed in the context of Industry 4.0, the so-called fourth industrial revolution. A key tenet of Industry 4.0 is the creation of smart factories where cyber-physical systems and the Industrial Internet of Things (IIoT) connect machines, products, and people in a dynamic value chain. CPS are "the key element in the implementation of the concepts of Industry 4.0"[54]. By ensuring that machines, work-in-progress items, and even logistics systems can directly communicate and coordinate, manufacturers aim to achieve higher autonomy and efficiency[54]. Klaus Schwab of the World Economic Forum famously described this trend as the "Fourth Industrial Revolution," emphasising how CPS and related technologies would transform industrial operations[54].

It is essential to note the relationship between CPS and the Internet of Things (IoT). While the IoT/IIoT focuses on connecting devices to the internet for data exchange, CPS encompasses a broader integration of those connected devices with real-time control and physical processes.

In manufacturing, IoT devices (sensors, smart instruments) provide data. Still, a CPS uses that data in closed-loop control, not just monitoring a machine on the internet, but actively adjusting it. Thus, CPS can be seen as an evolution of IoT in industrial contexts: all CPS involve connected devices, but not all connected devices form a full CPS. For example, simply putting a sensor on a machine and getting cloud alerts is IIoT. Still, when that sensor data is fed into a control algorithm that immediately alters the machine's operation, you have a CPS. In summary, Industrial IoT provides the connectivity and data streams that CPS in digital manufacturing leverage for real-time decision-making.

Digital Twins and Simulation

Another aspect of CPS integration is the use of digital twins – virtual models of physical assets. CPS often enable the creation of digital twins by constantly synchronising physical state data with a digital model[55]. In digital manufacturing, engineers can simulate production changes or identify optimisations in the twin before implementing them physically. The CPS ensures that the digital model stays up-to-date and that control actions from the digital side are executed in the physical world. This tight coupling significantly transforms how manufacturing systems are designed and operated, allowing for simulation-based design and optimisation in real-time[55].

Overall, CPS integration transforms digital manufacturing by providing real-time visibility and control over the production process. Physical production is no longer a black box executed by machines in isolation; instead, it becomes an agile, software-steered operation. This lays the groundwork for numerous efficiency, flexibility, and intelligence benefits, as discussed next.

Key Benefits of CPS in Digital Manufacturing

Implementing cyber-physical systems in manufacturing offers a host of benefits that drive performance, quality, and responsiveness. By merging real-time data analytics with physical control, CPS-based digital manufacturing systems achieve capabilities far beyond traditional automation. Below, we outline the key benefits, with examples, that CPS brings to the industry:

Improved Efficiency and Productivity: CPS greatly enhance production efficiency by optimising operations in real time. Automation guided by live data can reduce machine idle times, eliminate bottlenecks, and increase throughput. For example, sensors throughout a production line can feed data to scheduling algorithms that adjust the pace of upstream or downstream processes to balance the line and avoid delays[56]. One report notes that connecting sensors and actuators in CPS "improves manufacturing efficiency and flexibility" as part of the shift toward smart factories in the Industry 4.0 era[57]. Manufacturers like Bosch have reported higher production efficiency and reduced downtime after implementing CPS in their factories[56]. By automating repetitive tasks and coordinating equipment usage, CPS can significantly boost output per unit time without sacrificing quality.

Flexibility and Adaptability: Cyber-physical systems enable a level of agility in manufacturing that is hard to achieve with rigid, traditional machines. Because CPS devices are software-controlled and networked, reconfiguring a production process (for a new product variant or custom order) often means updating software or sending new instructions, rather than physically retooling machines. This adaptability lets factories respond faster to changes in demand or product design. For instance, a CPS-enabled assembly line can quickly switch its parameters for a different model of product by downloading a new configuration. At the same time, live sensor feedback ensures the new setup works correctly. Studies highlight that CPS allow factories to achieve remarkable adaptability in production processes, enabling, for example, rapid changeovers and mass customisation without significant downtime[58]. This flexibility extends to handling disturbances: if one machine in a network goes down, a CPS can reroute tasks to other machines or adjust workflows dynamically.

Real-Time Monitoring and Decision-Making: A hallmark of CPS is their ability to perform continuous real-time monitoring of equipment and product state, leading to informed decision-making on the fly. In digital manufacturing, every critical process parameter (temperature, pressure, speed, etc.) can be tracked by the CPS. The immense data streams are analysed by edge or cloud computing systems, which then can issue immediate control actions or recommendations. This closes the loop between monitoring and control in real time. The benefit is twofold: first, any anomalies or deviations can be caught instantly, preventing defects or equipment damage; second, opportunities for optimisation can be seized immediately. For example, if sensor data indicates a slight drift in a machining process, the CPS might automatically adjust the tool position to correct it within milliseconds. Real-time dashboards provided by CPS also help human operators make better decisions – they see up-to-the-second production metrics and can intervene with complete information. Overall, CPS give manufacturing systems something akin to a continuous "sense-think-act" capability, dramatically enhancing responsiveness and minimising latency in decision cycles[56, 59].

Higher Product Quality and Consistency: CPS help maintain tight quality control with fine-grained monitoring and control. They can detect quality issues early and automatically correct them or alert operators. For instance, in a CPS-enabled production line, if a product's dimension is trending out of tolerance, the system might adjust machine calibrations or segregate the item for inspection. Early defect detection and correction mean fewer off-spec products and less waste. Manufacturing data can be traced back for each item (genealogy), so root causes of quality issues are identified quickly. Indeed, manufacturers leveraging CPS have seen better quality outcomes, as real-time data allows them to catch deviations immediately[56]. One example is automotive painting processes: a CPS can monitor paint thickness and environmental conditions in real time and adjust spray parameters to ensure every car has a uniform coat, reducing rework. The net result is more consistent production and improved customer satisfaction due to reliable product quality.

Predictive Maintenance and Reduced Downtime: CPS in digital manufacturing significantly improve maintenance strategies by enabling predictive maintenance. Traditional factories often rely on periodic inspections or run-to-failure approaches, which can either be inefficient or risk unplanned breakdowns. In a CPS, machines continually report their condition (vibration, temperature, error codes, etc.). Advanced analytics can detect patterns indicative of impending failures – for example, a vibration signature suggesting a bearing is wearing out. Maintenance can then be scheduled just in time before a breakdown occurs. This predictive approach minimises unplanned downtime and extends equipment life. As an example, a network of CNC machines might each have a CPS module tracking spindle vibration; if one starts to show abnormal patterns, the system schedules a maintenance check during the next planned stop, avoiding a sudden crash. Real-world impact: General Electric's aviation division used CPS to monitor the production of jet engines, which helped reduce production time and cost by predicting issues before they halted the line[56]. Overall, CPS-driven predictive maintenance can dramatically raise asset uptime and maintenance efficiency.

Enhanced Safety for Workers and Equipment: CPS can improve safety in manufacturing environments by automating dangerous tasks and monitoring conditions. Robots and machines that are CPS-enabled can detect human presence and react (slowing or stopping to prevent accidents). They also monitor environmental factors (like gas levels, heat, and structural vibrations) and can trigger alarms or shutdowns to avert disasters. Because CPS often function autonomously, it can take workers out of hazardous process steps. For example, if a particular chemical process is risky, a CPS can automate the material handling while human workers supervise from a safe control room. Additionally, real-time alerts from CPS about any unsafe condition (an overheating motor, a pressure build-up, etc.) allow for quick responses. These features lead to safer operations, with fewer injuries and accidents. In essence, CPS act as an ever-vigilant guardian of the manufacturing process, ensuring it stays within safe bounds. Studies have noted safety as a key benefit, as CPS "enhanced monitoring and control of physical processes to prevent accidents"[59]. This is crucial in industries like aerospace or pharmaceuticals, where safety and compliance are paramount.

Better Resource Utilisation and Supply Chain Integration: Cyber-physical systems can also contribute to optimal use of resources (materials, energy) by smart scheduling and process optimisation. For instance, a CPS might adjust machine power settings when idle or coordinate heating/cooling cycles in an energy-intensive process to off-peak hours, thus saving energy. They can reduce material waste by detecting defects early or fine-tuning processes to use just enough material. Furthermore, because CPS links factory floor data to higher-level IT systems, it improves supply chain visibility. Real-time production data can be shared with suppliers and inventory systems, enabling just-in-time deliveries and reducing buffer stocks[56]. A CPS-enabled factory can, for example, automatically trigger a reorder of raw materials when sensors detect that current stock will be consumed in the next few hours of production, all without human intervention. This tight integration leads to leaner operations and cost savings across the value chain.

The integration of CPS in digital manufacturing yields a more efficient, agile, and intelligent production system. Companies implementing CPS have reported significant gains, from productivity improvements and quality gains to cost reductions and faster responsiveness to market changes[54, 56]. These benefits underscore why CPS are considered a game-changer for modern industry.

CPS Benefit	Example in Manufacturing	Outcome
Efficiency & Throughput	Real-time production scheduling adjustments on a CPS-enabled assembly line[56].	Higher machine utilisation; ~10% output increase (hypothetical).
Flexibility & Agility	Software reconfiguration of a robot cell for a new product variant.	Quick changeovers; minimal downtime between product runs.
Quality Improvement	In-process sensing and automatic tool correction in machining.	Defects are detected and corrected

		immediately, reducing the scrap rate.
Predictive Maintenance	Vibration monitoring on motors triggers bearing replacement before failure[57].	Fewer unexpected breakdowns; increased equipment uptime.
Safety	CPS-controlled robotic welder slows when a human worker approaches.	Reduced workplace injuries; safe human-robot collaboration.

Table: Summary of CPS benefits alongside examples of their manifestation in manufacturing settings (Note: Outcomes are illustrative; actual results vary by implementation)

Challenges and Limitations in Implementing CPS

While the benefits of CPS in manufacturing are compelling, deploying these systems is not without significant challenges. Industry professionals must navigate technical, organisational, and economic hurdles when implementing cyber-physical systems. Below are the important challenges and limitations associated with CPS in digital manufacturing:

High Implementation and Integration Costs: Building a CPS-ready manufacturing operation often requires substantial upfront investment. Legacy equipment might need to be retrofitted with sensors and connectivity, or replaced entirely with "smart" machines. Additionally, the software, networking infrastructure, and data analytics capabilities required for CPS can be expensive to develop and integrate. According to industry analysis, implementing CPS demands significant investments in hardware, software, and networking technologies[56]. For example, outfitting a factory with hundreds of IoT sensors and industrial Wi-Fi or 5G networks, plus the cloud computing resources to process the data, represents a considerable capital expenditure. For small and medium manufacturers, these costs can be prohibitive. Even larger firms must carefully calculate the return on investment and may roll out CPS in phases to manage expenses. Furthermore, maintenance of these advanced systems (software updates, sensor replacements, etc.) adds ongoing expenses. The challenge is to demonstrate that the long-term efficiency gains will outweigh the initial and recurring costs, which is not always straightforward.

Technical Complexity and Expertise Gap: CPS are inherently complex, involving multiple disciplines – from embedded systems and control engineering to IT and data science. Developing and maintaining a CPS thus requires a high level of specialised knowledge. Many manufacturers struggle with a skills gap: their workforce and engineering teams might not yet have experience with IoT architectures, machine learning, or cybersecurity, all of which can be part of a CPS implementation. One source notes that CPS implementation is "highly technical... requiring specialised skills and knowledge to implement and maintain"[56]. This complexity extends to system integration, making diverse devices and software interoperate smoothly can be very challenging. For instance, achieving real-time control loops that span cloud analytics and factory-floor actuators is complex and may require new system architectures. Debugging such systems is also non-trivial: problems can arise in the physical domain or cyber domain, or, worst, in the interaction between the two. Thus, companies often need to invest in training or hire new talent (data scientists, IoT architects, control engineers) to deploy CPS successfully. The multidisciplinary nature of CPS development can slow down projects and increase risk if not managed properly.

Interoperability and Legacy System Integration: A typical manufacturing CPS comprises many components – sensors from various vendors, machines of different vintages, PLCs running proprietary protocols, and so on. Ensuring that all these pieces communicate and work together seamlessly is a significant challenge. Interoperability issues arise because of a lack of standardisation or incompatible standards. As one analysis points out, a CPS often includes technologies and devices from different vendors, making seamless integration difficult[56]. For example, one machine might output data in a protocol that the rest of the system doesn't natively understand. Converting and standardising data from multiple sources (the OT – Operational Technology – systems) to feed into IT systems is a non-trivial task. Legacy equipment, which may not have any digital interface at all, poses another problem – they might require physical retrofitting or the use of external sensor kits to be included in the CPS, which can be clunky or limited. Industry consortia and standards bodies are addressing interoperability by promoting common standards (such as OPC UA for industrial data exchange, or MQTT/REST for IIoT communication). However, in practice, many factories still encounter integration pains. This challenge means that CPS projects often start with pilot programs on a limited set of machines before scaling up, tackling integration one step at a time.

Cybersecurity Threats and Privacy: Because CPS connect physical machinery to networks (often corporate networks or even the internet), they become targets for cyberattacks. Cybersecurity is a top concern in CPS implementation[56]. A breach or malware in a CPS isn't just an IT issue – it can disrupt physical operations, damage equipment, or even endanger workers. For instance, a hacker who gains access to a factory's CPS could potentially manipulate sensor readings or machine instructions, causing malfunctions. The infamous case of the Stuxnet worm (which targeted industrial control systems) is a reminder of how cyber attacks can physically sabotage production. Manufacturers, therefore, must invest in robust cybersecurity measures for CPS: network segmentation, encryption of sensor data, authentication for devices, regular security updates, and intrusion detection systems tailored to industrial environments. This is challenging because many legacy OT devices were not designed with security in mind and may have vulnerabilities. Additionally, more connectivity means a larger attack surface – every sensor or smart device added is a potential ingress point if not properly secured. Alongside cybersecurity, data privacy can be a consideration, especially if CPS data is shared across corporate boundaries or with cloud services. Companies need to ensure sensitive production data or intellectual property is protected in transit and at rest. Addressing these security and privacy issues is critical;

failure to do so can negate many of the benefits of CPS by introducing new risks.

Reliability, Safety, and Trustworthiness: A CPS in manufacturing often controls mission-critical operations – its correct functioning is essential to avoid production loss or accidents. This raises the bar for reliability and safety in system design. Any latency or downtime in the cyber components (like a slow network or a software crash) could have immediate physical consequences, such as a machine collision or product scrap. Therefore, CPS implementations must be rigorously engineered for dependability. Some CPS require "dependable, high-confidence or provable behaviours"[53], meaning they need to be validated and sometimes formally verified to ensure they behave correctly under all expected conditions. Achieving this can be difficult given the complexity of the system and the unpredictable nature of real-world environments. Real-time requirements are also challenging – if a CPS relies on cloud computing, what happens if the network lags? Manufacturers may need to incorporate edge computing to keep critical control loops local and fast. Moreover, ensuring safety in a system that can act autonomously means implementing proper fail-safes and overrides. For example, if sensors conflict or a piece of the system goes offline, the CPS should fail gracefully (perhaps reverting to manual control or a safe shutdown). Building this kind of resilience and fault-tolerance adds complexity. Gaining the trust of engineers and operators in an autonomous CPS is also a hurdle. There can be reluctance to cede control to an algorithm, especially in sectors with a strong safety culture (like aerospace). Hence, extensive testing, validation, and phased rollouts are necessary to build confidence in CPS operations.

Organisational and Workforce Challenges: Introducing CPS often requires changes in workflows and job roles. Workers on the factory floor need to adapt to new interfaces and ways of working. There may be resistance to change, especially if the CPS introduction is seen as a threat to jobs (automation replacing specific tasks). Workforce training is essential so that employees can operate and maintain CPS equipment[56]. For instance, maintenance technicians need to learn to interpret data analytics dashboards in addition to turning wrenches.

There is also a cultural shift toward data-driven decision making that organisations must embrace; production managers must learn to trust insights from CPS analytics. Companies might face initial drops in productivity as staff climb the learning curve of new CPS tools or as processes are re-engineered to exploit CPS capabilities. Effective change management, involving operators in the design and rollout of CPS (to get buy-in and leverage their expertise), can mitigate this. In addition, new standards and procedures may be needed – for example, how to handle software updates on machines, or new safety protocols for human-robot interaction. All these organisational factors can slow down CPS adoption if not proactively addressed.

Despite these challenges, industry trends show a steady march toward CPS implementation. Overcoming the hurdles often involves cross-disciplinary collaboration (IT with OT teams, hiring cybersecurity experts, etc.), adopting standards, and sometimes reaching out to external solution providers for support. In many cases, incremental adoption is key: starting with pilot projects that demonstrate value, then scaling up while learning and addressing challenges in stages, rather than attempting a big-bang deployment.

Use Cases and Sector Examples

Cyber-Physical Systems are being applied across various manufacturing sectors, revolutionising how products are made. Below are several sector-specific use cases and case studies illustrating CPS in action:

Automotive Manufacturing

The automotive industry has been at the forefront of adopting CPS as part of the drive toward smart factories. Modern car production involves highly automated assembly lines where hundreds of robots, machines, and conveyors must work in unison – a perfect scenario for CPS integration. In a CPS-enabled automotive plant, each robot welder, paint sprayer, or assembly station is equipped with sensors and networked controllers. These components communicate their status (position, task completion, any faults) to a central system and each other, coordinating the complex choreography of vehicle assembly.

Real-time Production Optimisation: BMW, for instance, has implemented CPS in its production processes to enable real-time data exchange and analysis across its factories. This connectivity allowed BMW to optimise production processes, reduce waste, and improve product quality on the fly[56]. If one part of the line is moving slower, the upstream processes can automatically decelerate to match pace, preventing bottlenecks or overproduction. Data analytics might reveal inefficiencies in the workflow that engineers can address, such as redistributing tasks among robots to balance their load. The result is a smoother production flow and higher throughput.

Smart Robotics and AGVs: Automotive plants also utilise Autonomous Guided Vehicles (AGVs) or Autonomous Mobile Robots (AMRs) to deliver parts to the line. When these vehicles operate as part of a CPS, they can adjust their routes in real time based on production needs – e.g. if a particular station is starved of parts, the CPS dispatches an AGV there immediately. This kind of responsive logistics is only possible when all elements (stations, vehicles, inventory systems) are interconnected and share data continuously.

Quality Control: Another CPS application in automotive is in quality inspection. Traditionally, end-of-line inspection was manual or sampling-based. Now, manufacturers like Toyota and Volkswagen employ CPS with machine vision cameras and sensors at various points in production. These systems automatically detect defects (like a missing bolt or a paint blemish) and either correct them or flag the vehicle for rework. The data from these inspections is fed back into the production CPS to identify where in the process the issue arose, enabling continuous improvement.

Automotive case studies have shown significant gains from CPS. Bosch reported increased production efficiency and reduced downtime after retrofitting their lines with CPS and IoT connectivity[56]. General Motors has used CPS concepts in implementing flexible manufacturing systems that can build multiple car models intermixed, adjusting robot programs in real time as different car bodies move down the line. Overall, CPS helps the automotive sector achieve the holy grails of manufacturing: high volume, high quality, and high flexibility, all at once.

Aerospace and Defence Manufacturing

Aerospace manufacturing involves complex, low-volume, and high-precision processes, such as building jet engines, aircraft assemblies, or defence equipment. The cost of errors or downtime is extremely high in this sector, and the quality requirements are stringent. CPS technologies are being used to enhance both the efficiency and reliability of aerospace production.

Engine Manufacturing (GE Aviation): Building a jet engine involves many steps (casting, machining, assembly, testing) and each engine is serialised and tracked. GE Aviation implemented CPS in its manufacturing lines to achieve real-time monitoring and analysis of production processes[56]. Sensors on tools and machines gather data, such as torque, vibration, and temperature, during engine component fabrication. By analysing this data, GE's systems can detect subtle signs of tool wear or process drift that might affect the engine's quality. They reportedly reduced the time and cost of producing engines by using this real-time data to preempt problems and maintain optimal process settings[56]. For instance, if a drilling operation in a turbine blade were starting to deviate from the centre, the CPS would detect it through sensor feedback and adjust or call for maintenance before a batch of blades is produced incorrectly. The result is fewer defects and rework, as well as a faster overall production cycle.

Aircraft Assembly: In large aircraft assembly factories (such as Boeing's or Airbus's), CPS are used to coordinate tasks, including automated drilling robots that work on fuselages. These robots must work in precise synchronisation and often in tandem with human workers (augmented by wearable CPS devices for safety). A CPS network can manage the operation of multiple robots on an airframe, ensuring they don't collide and that each drilling operation is performed at the exact specified location and angle. All the data from these drilling processes is recorded (forming a digital twin of the aircraft under assembly), which is valuable for later maintenance and quality assurance. Deviations can be traced and addressed immediately.

Aerospace is pioneering advanced techniques, such as 3D printing (additive manufacturing), for creating complex parts. CPS comes into play by monitoring the additive process layer by layer with sensors (such as cameras and thermal sensors) and utilising AI algorithms to detect defects during the build. For critical components like fuel nozzles, a CPS might halt a print if it detects a flaw, saving time and materials compared to discovering an issue in post-production inspection. Moreover, CPS can integrate non-destructive testing equipment (such as ultrasound or X-ray inspection machines) with production data, automatically linking quality data to the specific process parameters used, thereby closing the loop for process improvement.

Defence Manufacturing: In defence electronics or equipment manufacturing, which often takes place in secure and controlled environments, CPS are utilised for environmental monitoring and asset tracking. For example, a missile assembly facility may utilise CPS to continuously monitor humidity, dust, and other environmental parameters, as these can impact sensitive electronics. The CPS can automatically adjust air filtration or alert managers if conditions drift out of spec, ensuring product integrity.

In summary, aerospace and defence sectors leverage CPS for precision, traceability, and predictive insight. By capturing vast amounts of data and controlling complex processes in real-time, CPS help these sectors maintain the highest quality standards while seeking improvements in efficiency for very sophisticated products.

Precision and Electronics Manufacturing

Precision manufacturing refers to industries that require extremely tight tolerances and often deal with small-scale or high-accuracy production, such as semiconductor fabrication, electronics assembly, medical device manufacturing, and advanced tooling. In these areas, CPS is a critical enabler for maintaining precision and yield.

Semiconductor fabs are arguably already cyber-physical systems, with hundreds of tools (lithography machines, etchers, deposition equipment) coordinated by manufacturing execution systems. CPS takes this further by adding more sensors and advanced control. For instance, semiconductor equipment may have sensors monitoring particle contamination, and a CPS can dynamically adjust air flow or schedule maintenance when particle counts rise. Advanced Process Control (APC) in semiconductor manufacturing is a form of CPS, where equipment adjusts recipe parameters in real-time based on measurements from metrology tools on recently processed wafers. This closed-loop control helps maintain critical dimensions and yields within target values. Given the high cost of wafer defects, these CPS controls can save millions by improving yield percentages even marginally. Moreover, predictive maintenance is crucial in fabs – unplanned tool downtime can halt production lines, resulting in thousands of dollars in lost revenue per minute. CPS solutions that predict wear of parts (such as a sensor detecting vibration in a wafer handler robot) enable fabs to replace components during scheduled maintenance windows, thereby avoiding costly downtime and protecting throughput.

Electronics Assembly (PCB manufacturing): In printed circuit board (PCB) assembly for electronics, CPS is used to link pick-and-place machines, solder reflow ovens, and automated optical inspection (AOI) systems. The CPS can track each board through the line, and if AOI finds a solder defect or misaligned component at the end, the system can trace it back to the exact time and settings of the soldering oven or placement machine that handled it. If a trend is noticed (say, one particular placement head is misaligning parts), the CPS will flag it for correction. This ensures high quality and yields for tiny components (where human inspection is nearly impossible). Additionally, these factories employ CPS for flexibility – rapidly changing production from one PCB design to another by downloading new machine programs, guided by a central system that coordinates the changeover and verifies that each machine's setup (feeder positions, solder profiles) matches the new product's requirements.

Medical devices often require precision and strict compliance with regulations. CPS can help maintain audit trails and process control. For instance, in manufacturing a stent or prosthetic, a CPS might record every environmental condition and machine setting used during production. If any parameter exceeds the validated range, the CPS halts the process and alerts engineers, ensuring that no potentially non-compliant product is released. In highly automated pharma or biotech production (which is process manufacturing but precision-oriented), CPS and digital twins are used to maintain product quality. One case utilises a CPS with a digital twin to control a bioreactor for drug manufacturing – sensors feed cell growth data to a model that adjusts nutrient flow or temperature in real-time to keep the culture on track [60, 61]. Such precise control can significantly improve the yield of biological products that are highly sensitive to environmental conditions.

Across precision and electronics manufacturing, CPS provides the confidence and control needed to hit very tight tolerances consistently and to document that control (for regulatory or quality purposes). Companies in these sectors also benefit from speed, as they can quickly ramp up new products by simulating and fine-tuning processes virtually with CPS before committing them to physical production.

Additional Use Case – Process Industries

(While the focus is on digital manufacturing (discrete products), it's worth noting CPS use in process industries like chemicals, oil & gas, or food processing, as they face similar integration challenges. For example, a refinery might use CPS to convert traditional tank farms into smart systems with remote monitoring, enabling predictive maintenance on valves and pumps instead of manual checks[54]. These applications underline that CPS principles apply broadly anywhere physical processes can be digitally monitored and controlled.)

Each of these examples demonstrates how CPS can be tailored to the specific needs of a sector, whether it's achieving ultra-precision, ensuring traceability, or managing complexity. Importantly, the successes in these use cases are driving wider adoption: as one factory in a sector demonstrates the benefits of CPS, others quickly follow to stay competitive. As a result, CPS are becoming the new normal in cutting-edge manufacturing settings, from car plants to chip fabs.

Standards, Frameworks, and the Road Ahead

The rise of cyber-physical systems in manufacturing has prompted efforts to develop standards and frameworks that guide their implementation, ensuring interoperability and security. Industry professionals looking to adopt CPS should be aware of ongoing standardisation initiatives and how CPS fits into larger strategic frameworks, such as Industry 4.0 and IIoT.

Industry 4.0 Reference Architectures: As mentioned, CPS is a foundational concept in Industry 4.0. In Germany, the Reference Architectural Model for Industry 4.0 (RAMI 4.0) was developed, in which CPS (often termed "cyber-physical production systems") play a central role in connecting the enterprise IT level with the shop-floor device level. This model, along with concepts from the Industrial Internet Consortium (IIC) in the United States (which coined the term Industrial Internet), provides high-level guidance on integrating CPS with business processes. The common theme is that standardised communication and data models are needed so that devices can seamlessly join a CPS. OPC UA (Open Platform Communications Unified Architecture) is one widely adopted standard protocol for machine data in manufacturing that aligns with these visions, enabling different vendors' equipment to communicate in a CPS setting.

NIST CPS Framework: The U.S. National Institute of Standards and Technology (NIST) has been actively involved in CPS standardisation. NIST Special Publication 1500-201, "Framework for Cyber-Physical Systems", provides an organised methodology to analyse and engineer CPS[54]. It outlines facets such as functional domains (e.g., sensing, control, communication), aspects like security and reliability, and cross-cutting concerns. By following a framework like NIST's, organisations can ensure they consider all critical elements of CPS design (from interoperability to security to timing constraints). NIST's work often emphasises "trustworthiness" of CPS – covering safety, security, privacy, resilience and reliability – recognising that these are vital for adoption. For manufacturers, aligning with such frameworks can help in designing CPS that meet regulatory and best-practice guidelines. For instance, NIST's Cybersecurity Framework has a manufacturing profile (NIST SP 800-82) that, while focused on control systems security, complements CPS deployment by highlighting how to protect these connected systems[62].

ISO and IEC Standards: Internationally, ISO/IEC are also developing CPS-related standards. ISO has recognised the need for common definitions and models for CPS[63]. One example is the ISO/IEC AWI 5689 project, which is working on a technical specification for a CPS conceptual model and security frameworks[64]. Additionally, existing relevant standards are often applied to CPS implementations. For instance, ISO 26262 (functional safety for road vehicles) becomes highly relevant when implementing CPS in automotive, to ensure that the integration of software and hardware meets safety requirements[65]. In manufacturing, IEC 62443 standards (for industrial control system security) are frequently referenced to secure CPS, since they provide best practices for protecting control networks. We also see efforts, such as SAE's G-32 committee, developing guidelines for CPS security in aerospace and automotive contexts.

Industrial IoT Platforms and Consortia: Many CPS in manufacturing are built upon Industrial IoT platforms provided by major automation vendors (Siemens, ABB, Rockwell, etc.) or tech companies. These platforms often incorporate de facto standards and reference architectures. For example, the Industrie 4.0 Platform in Germany (a consortium) publishes guidelines on implementing CPS/IIoT in line with RAMI 4.0. The IIC's Industrial Internet Reference Architecture (IIRA) similarly offers a blueprint. While not formal standards, these frameworks aim to align the industry on common approaches, enabling ecosystems of devices and software to work together in a CPS environment.

Relationship to IIoT and Cloud Standards: CPS doesn't exist in isolation; it overlaps with IoT and cloud computing. Thus, standards like MQTT (a lightweight publish/subscribe protocol widely used for

Industrial Internet of Things, IIoT) or OPC UA over TSN (Time-Sensitive Networking, for ensuring the timely delivery of messages in a network) are essential pieces for real-time CPS data communication. IEEE is working on standards for time synchronisation (like IEEE 1588 Precision Time Protocol), which is crucial for CPS that require coordinated timing between devices (e.g., synchronised robot motion). All these technical standards contribute to making CPS more plug-and-play and reliable.

For professionals in digital manufacturing, staying up-to-date with these standards is crucial for future-proofing their CPS investments. Adhering to widely adopted standards can ease integration (solving the interoperability challenge) and ensure compliance with safety and security expectations. It also avoids vendor lock-in by allowing mixing and matching of components in the CPS.

Future Outlook: As we look ahead, CPS in manufacturing will continue to evolve. Emerging technologies, such as 5G and 6G wireless, are expected to further enable CPS by providing ultra-reliable, low-latency communication on the factory floor, making wireless sensor-actuator loops feasible at scale. Artificial intelligence and machine learning will be increasingly embedded in CPS (sometimes referred to as "smart CPS"), bringing more advanced predictive analytics and even autonomous decision-making to the production line. CPS and AI are converging to create self-optimising production systems that learn and improve over time. Standards and best practices will also mature around AI in CPS (for example, ensuring the AI decisions are transparent and safe).

Another concept on the horizon is greater human-CPS collaboration. Rather than fully autonomous lights-out factories, many envision CPS that work in tandem with skilled workers, amplifying human capabilities. This requires designing CPS with intuitive human-machine interfaces and considering ergonomic and human-factor standards as well.

In conclusion, Cyber-Physical Systems are transforming digital manufacturing by bridging the divide between the physical factory and the digital enterprise. They bring unprecedented levels of control, adaptability, and insight to production operations, enabling the vision of smart, connected factories under the Industry 4.0 paradigm. While challenges in implementation exist – from technical complexity to security – the progress in standards and successful case studies across automotive, aerospace, precision manufacturing and beyond are paving the way for broader adoption. Industry professionals who master CPS technologies and principles will be at the forefront of driving efficiency, innovation, and competitiveness in manufacturing for years to come. With CPS, the factories of the future are becoming a present reality – where machines not only toil, but think, communicate, and continuously improve the way things are made, heralding a new era of industrial productivity and creativity.

Cloud and Edge Computing in Manufacturing

Manufacturing is undergoing a digital transformation as part of the Industry 4.0 era, characterised by the convergence of information technology (IT) with operational technology (OT) on the factory floor[66]. In practice, this means that traditionally isolated production machines and control systems are now connected to networks, feeding data into advanced analytics and automation systems. Cloud computing and edge computing are two pivotal technologies enabling this transformation. They provide the backbone for smart factories by handling massive data from sensors and machines, powering real-time decision-making, and enhancing transparency across operations[66]. Manufacturers today face pressures to reduce downtime, improve efficiency, and increase supply chain resilience. By leveraging cloud and edge computing, they can merge enterprise IT capabilities with shop-floor OT data to meet these goals with greater agility and insight. This section defines cloud and edge computing in a manufacturing context, explores their applications and benefits (from predictive maintenance to remote monitoring), examines the challenges (such as latency and security), and reviews real-world examples. It also discusses hybrid cloud-edge strategies and highlights the standards and emerging trends that are shaping the future of manufacturing.

Cloud Computing in Manufacturing

Cloud computing refers to the delivery of computing services (such as servers, storage, databases, software, and analytics) over the internet on an on-demand basis. Rather than hosting applications or storing data on local servers, companies use remote data centres operated by cloud providers. In manufacturing, cloud computing has become an essential enabler of modern operations[67]. It allows manufacturers to offload heavy computational tasks and data storage to centralised servers, accessible from anywhere. For example, production data, machine logs, and quality metrics can be continuously uploaded to cloud platforms for analysis and review. Cloud platforms (such as Amazon Web Services, Microsoft Azure, and Google Cloud) offer vast, scalable resources to run advanced manufacturing applications, ranging from enterprise resource planning (ERP) to data analytics and artificial intelligence.

A key characteristic of cloud computing is centralisation. Data from one or many facilities is transmitted to a central cloud data centre for processing. This centralisation simplifies remote access and global visibility. All devices or users that need to access specific data or applications connect to the cloud server, ensuring a "single source of truth" for distributed operations[68]. For instance, a cloud-based Manufacturing Execution System can compile production reports from multiple plants, and managers anywhere can view real-time dashboards through a web interface. Because everything is in one place, scalability is straightforward – if more storage or computing power is needed, it can be provisioned on the cloud without on-site hardware upgrades. Similarly, maintenance and security of the infrastructure are primarily handled by the cloud provider, which often means stronger security measures than a small IT team could implement locally (such as robust data encryption, access controls, and continuous updates)[67]. Manufacturers also benefit from a pay-as-you-go model: instead of significant capital expenditures on servers, they pay for cloud resources as operational expenses, only for what they use, which reduces IT costs[67].

In manufacturing contexts, cloud computing supports a wide range of applications. It can host enterprise software for supply chain management, inventory control, and product lifecycle management, integrating these with shop-floor data[67]. Cloud-based IoT (Industrial Internet of Things) platforms aggregate data from sensors and machines across the factory network. For example, environmental sensors might send temperature and humidity readings to a cloud database, and production equipment controllers might log cycle times or error codes to the cloud for analytics. By tracking production data, material usage, quality metrics, and more, cloud platforms give manufacturers a holistic view of their operations in real time[67]. This remote visibility extends beyond a single site – a plant manager or engineer can monitor multiple facilities around the world through cloud dashboards, receiving instant alerts if any parameter exceeds its range. Such remote monitoring and control capabilities were previously difficult with on-premises systems[67]. Additionally, the cloud's virtually unlimited storage and compute power enable big data analytics on manufacturing information. Complex algorithms can run in the cloud to find patterns in production data, fueling insights for process improvement and predictive models. In short, cloud computing in manufacturing provides a scalable, flexible platform to unify data and applications across the enterprise, breaking down silos between departments and sites for greater efficiency.

Edge Computing in Manufacturing

Edge computing brings computation and data storage closer to the devices and processes in the factory, rather than relying solely on a distant cloud server. In essence, an edge computing device (such as an industrial PC, IoT gateway, or even a smart sensor or PLC with processing capability) is deployed on-site, at the "edge" of the network, near the machines. Its role is to collect, filter, and analyse data locally and even make automated decisions in real-time, without needing to send everything to the cloud first. In manufacturing, the goal of edge computing is to process and analyse data near a machine that needs to act on that data in a time-sensitive manner[68]. By doing so, it minimises the latency (delay) of data transmission and avoids network bottlenecks. For example, imagine a factory sensor detecting that a machine tool's vibration exceeds a critical threshold. If this data had to travel to a cloud data centre hundreds or thousands of kilometres away for analysis, even a few hundred milliseconds of delay could be too late to prevent a breakdown. Edge computing enables sensor data to be processed on a nearby edge device immediately, triggering an alarm or an automatic machine shutdown within milliseconds.

The defining characteristics of edge computing are decentralisation and real-time responsiveness. Instead of funnelling all raw data to a single cloud repository, edge devices handle data at or near its source. As a

Red Hat overview describes, edge computing (as opposed to pure cloud computing) allows manufacturers to perform automation and machine-to-machine communication "closer to the source, rather than sending data to a server for analysis and response"[66]. For instance, advanced industrial robots or machine vision systems on an assembly line might be connected to a local edge controller that runs the AI models for object detection or quality inspection on-site. If a defect is spotted on a product, the edge device can immediately reject that item or adjust the robot's parameters, achieving response times on the order of milliseconds. This is crucial for use cases like automated quality control, safety interlocks, or precision adjustments in process parameters, where any delay could result in defective products or equipment damage. By processing data locally, edge computing also dramatically reduces the volume of data that must be sent over networks. Modern equipment can generate enormous data streams (a single machine tool can produce hundreds of data points every millisecond)[68]. Transmitting every bit of that to the cloud is inefficient and can overwhelm network bandwidth. Edge devices can pre-process and summarise data (for example, calculating averages, detecting anomalies, or filtering out irrelevant readings) and send only the meaningful results to the cloud. This optimises bandwidth usage and ensures the cloud is not clogged with unnecessary detail[68].

Another benefit of keeping certain computations at the edge is enhanced reliability and autonomy. Manufacturing sites cannot always depend on continuous high-speed connectivity to a remote cloud. If the network link goes down or becomes slow, cloud-reliant systems might halt. Edge computing enables critical operations to continue locally even if the connection to the cloud is intermittent. It effectively provides a level of self-sufficiency on the factory floor for the most mission-critical control tasks. For example, a packaging line with an edge controller can continue to run with real-time adjustments even if the internet connection drops, and then sync aggregate data to the cloud when the connection is restored. In regulated or sensitive manufacturing (such as aerospace or medical device production), edge computing can also address data governance concerns by keeping proprietary or sensitive data on-site. Instead of streaming everything to third-party cloud servers, companies can choose to process sensitive information within their facility boundaries. This minimises exposure of data and can help meet data sovereignty or compliance requirements (69) (we will discuss this further under challenges). In summary, edge computing in manufacturing involves pushing intelligence to the shop floor, thereby localising decision-making, reducing latency, and enabling the factory to react to events instantaneously. It complements cloud computing by handling the immediacy of operations, while the cloud handles broader data analysis and coordination, as we will explore in hybrid approaches.

Cloud vs. Edge – A Comparison

In practice, cloud and edge computing are not mutually exclusive but serve different needs. Cloud computing centralises resources for global optimisation, heavy analytics, and cross-facility integration, whereas edge computing decentralises resources for speed, real-time control, and data reduction at source[68]. The following table highlights key differences and roles of each in manufacturing:

Aspect	Cloud Computing	Edge Computing
Location	Remote data centres (off-site, provided by cloud vendors).	On-site devices at the factory edge (on the shop floor or nearby).
Latency	Higher latency – data travels to central servers and back.	Ultra-low latency – processing happens close to machines, in real time.
Processing	Batch or large-scale processing, big data analytics, and historical trends[68].	Immediate processing of sensor/device data for instant action[68].
Use Cases	Multi-site data aggregation, long-term analytics, enterprise applications (e.g.	Real-time control, equipment monitoring, safety interlocks, and on-the-fly quality control [66, 69].

	production planning, supply chain)[67].	
Scalability	Virtually infinite scalability using cloud provider resources; easy to add storage/compute on demand[67].	Scales by deploying more local devices; can be limited by on-premises hardware footprint[66].
Network Dependence	Requires reliable network connectivity for continuous operation (mitigated by local buffering at times).	Can operate autonomously; less dependent on constant cloud connectivity for critical tasks.
Security & Governance	Centralised control can simplify security management; however, data leaves the factory and resides off-site (trust in provider needed)[67].	Data stays on-site, reducing external exposure[69], but requires securing many distributed devices and physical locations.

Table: Cloud vs Edge Computing

Both paradigms work in tandem in modern smart factories. Typically, raw data is generated at the edge, initial critical filtering and control are done locally, and then aggregated data or insights are sent to the cloud for deeper analysis and integration with business systems [70]. This synergy is further discussed in the hybrid models section.

Applications in Manufacturing Processes and Systems

Cloud and edge computing are being applied across the manufacturing value chain, from the shop floor processes to higher-level production management and supply chain systems. A typical smart manufacturing architecture might involve numerous IoT sensors and devices on equipment, edge gateways at each production line or cell, and cloud platforms that consolidate information enterprise-wide. This section outlines how these technologies embed into manufacturing processes and infrastructure:

Real-Time Production Monitoring: On a factory floor, machines such as CNC machines, presses, or robotic arms are equipped with sensors (measuring temperature, vibration, speed, etc.) and controllers that can connect to a network. Edge devices or industrial controllers collect this machine data in real time. For example, an edge IoT gateway might interface with a machine's PLC and stream data about cycle times and part counts. Instead of sending every sensor reading to the cloud, the gateway can locally compute key metrics, such as the machine's current utilisation or if its performance is deviating from the norm, and send summary data or alerts upstream. Meanwhile, a cloud-based Manufacturing Execution System (MES) or IoT platform receives the data (directly or via the edge device) and stores it for visualisation and analysis. Using a cloud IoT platform, manufacturers get real-time dashboards of machine status and performance across the shop floor[67]. This empowers production managers to see, for instance, which machines are down, which are starved of material, and overall equipment effectiveness (OEE) metrics from any location. If a machine's edge device detects an anomaly (e.g., a temperature spike), it can trigger an immediate local action (like adjusting cooling or pausing the machine) and simultaneously log the event to the cloud, where an engineer is notified via a remote monitoring application.

Process Control and Automation: Edge computing enables advanced process control strategies by hosting control logic closer to equipment. Consider a robotic welding station in an automotive assembly line. High-speed sensors and cameras monitor each weld's quality (temperature, alignment, etc.). By deploying an edge computer directly at the station, running machine vision and AI algorithms, the system can instantly analyse each weld seam. If the weld is suboptimal, the edge device might adjust the welding parameters on the next part or instruct the robot to rework the seam, all within milliseconds. Such tight feedback loops would be impossible if data had to travel to a cloud and back. In one example, a high-speed packaging line used edge computing to perform visual quality inspections in real time – products were inspected within milliseconds, and any defect triggered an immediate mechanical adjustment, which prevented defective packages without slowing down the line[69]. These kinds of edge-driven automation improve precision and reduce waste. The cloud complements these operations by later aggregating data from multiple cycles or stations to identify broader patterns. For instance, the cloud might reveal that weld quality issues spike at certain times of day or correlate with a particular batch of materials, insights that the on-site edge system alone wouldn't analyse.

Predictive Maintenance Systems: One of the most transformative applications is the use of cloud and edge technologies in tandem for predictive maintenance of equipment. Machines are outfitted with sensors that measure vibration, acoustics, temperature, power draw, and other indicators of machine health. An edge device on each machine (or a cell of machines) continuously analyses these sensor signals using algorithms (sometimes AI models) to detect anomalies that precede a failure. For example, an abnormal vibration signature may indicate that a bearing is starting to wear out. The edge device can flag this in real time and perhaps slow down the machine to prevent immediate damage. It also sends the data to a cloud-based maintenance application, where more intensive analysis is performed, comparing it against historical data and failure models. Over time, the cloud platform, which collects data from many machines and even multiple facilities, can predict when a machine is likely to fail and schedule maintenance proactively. This is already happening in many plants: assets on the shop floor create alert tickets and maintenance schedules automatically when edge analytics detects underperforming components[68]. For example, an edge module on a critical pump might notice increasing vibration and alert the cloud maintenance system, which then creates a maintenance work order for that pump in the company's central maintenance management software. This approach shifts maintenance from routine, calendar-based intervals to condition-based interventions, minimising unplanned downtime. The benefits include reduced machine downtime, lower maintenance costs, and improved equipment reliability[68]. Unplanned downtime is estimated to cost manufacturers tens of billions of dollars annually, so preventing failures yields a significant financial return [69].

Supply Chain and Production Planning: Cloud computing plays a crucial role in connecting manufacturing operations with supply chain systems. When factory data (inventory levels, production rates, quality yields) is pushed to the cloud, it can be shared with other enterprise systems or even partners. Manufacturers use cloud-based supply chain management tools to get end-to-end visibility. For instance, as soon as production output data is logged in the cloud, the system can trigger the automatic reorder of raw materials if inventory falls below a threshold, or update customers on expected delivery times. In the aerospace sector, companies are leveraging cloud connectivity to create digital supply networks – virtual representations of the supply chain that mirror each step from design to delivery[71]. By incorporating data from multiple ecosystem partners (suppliers, logistics providers, etc.) into a cloud platform, they can simulate and dynamically adjust to changes. Edge computing assists by providing timely data from the factory floor (e.g., actual production count vs. plan), which the cloud uses to adjust procurement or distribution. The result is a more agile supply chain that can adapt to disruptions and changes in demand[71]. For example, suppose a machine at a supplier's facility goes down. In that case, an edge device at that supplier might flag a delay, and the cloud system immediately alerts the OEM to reroute orders to a backup supplier, thus avoiding a line stoppage.

Facility Management and Energy Optimisation: Cloud-edge architectures are also used for managing the factory environment and utilities. Sensors around a plant monitor power consumption, HVAC (heating, ventilation, air conditioning) systems, and other utilities. Edge controllers can adjust thermostats or lighting in real-time based on occupancy or machine operation (for instance, reducing cooling when specific high-heat machines are not in use). All these facility data points are sent to cloud analytics, which optimise energy usage across the plant. Over time, analytics may reveal that specific machines can operate at off-peak times to conserve energy or that a particular building's climate control system can be optimised for greater efficiency. Cloud-based energy management systems in manufacturing have led to reduced energy costs and support corporate sustainability goals.

In summary, cloud and edge computing are deeply integrated into modern manufacturing operations. Edge computing provides immediate, on-site responsiveness, powering automation, control, and local data processing in processes such as assembly, machining, and inspection. Cloud computing offers the overarching intelligence and coordination, connecting operations to enterprise systems, performing large-scale analytics (such as analysing years of production data for trends), and enabling remote management. Together, they allow use cases ranging from a single machine's health monitoring to global production network optimisation, forming the technology foundation of intelligent manufacturing systems.

Key Benefits for Manufacturing Operations

The adoption of cloud and edge computing in manufacturing yields numerous benefits that directly impact operational performance, efficiency, and flexibility. Below are some of the key advantages for manufacturing operations, along with explanations of how each is achieved:

Predictive Maintenance and Reduced Downtime: One of the most celebrated benefits is the ability to predict and prevent machine failures before they happen. By continuously monitoring equipment through edge devices and analysing trends in the cloud, manufacturers can transition from reactive repairs to proactive maintenance. Predictive maintenance models identify early warning signs of wear or malfunction, such as an increase in motor vibration or temperature, and notify maintenance teams to address the issue during planned downtime. This significantly reduces unexpected breakdowns. The value is substantial: implementing data-driven predictive maintenance is shown to reduce unplanned downtime and maintenance costs while increasing equipment lifespan[68]. A study noted that unplanned downtime costs manufacturers around $50 billion annually[69], so even a slight percentage reduction has significant payback. Real-world implementations have reported dramatic improvements; for instance, a manufacturer using edge analytics on its machines achieved a 62% decrease in unplanned maintenance downtime after deploying predictive maintenance solutions[69]. Less downtime means higher production uptime and throughput, directly improving the bottom line.

Scalability and Flexibility of Operations: Cloud computing offers virtually unlimited scalability, a boon for growing manufacturers or those with fluctuating demand. In traditional setups, adding capacity (whether computing power for data analysis or new software capabilities) often required purchasing and installing new servers and IT infrastructure on-site. With cloud-based infrastructure, manufacturers can scale up or down on demand – for example, spinning up additional servers in the cloud to run a complex production simulation, or expanding storage to accommodate a surge in production data – without investing in new hardware themselves. This pay-as-you-go scalability also means manufacturers can experiment with new digital initiatives at lower risk, since they can start small and grow the cloud resources as needed. A company can quickly roll out a new analytics application to all its plants by deploying it in the cloud, rather than installing software at each site. Flexibility is enhanced because cloud services allow easy integration of new functionalities (like IoT services, AI tools, etc.) and support multi-site operations seamlessly. As an example, Polamer Precision, a precision aerospace parts manufacturer, chose a cloud-based manufacturing platform specifically for its potential to expand and adapt to the company's needs – the "cloud aspect" and easy customization of the system made the decision easy as the company anticipated growth and needed their IT systems to grow with them[72]. In short, cloud computing transforms infrastructure into a flexible resource that can match the pace of manufacturing business needs, whether scaling production, introducing new product lines, or expanding into new locations.

Remote Monitoring and Collaboration: By centralising data and applications, cloud computing enables remote access to manufacturing information from anywhere. This is hugely beneficial for global operations and for collaboration among teams. Plant managers and engineers no longer have to be on-site to know what is happening in production – they can check cloud dashboards for live metrics on throughput, quality, and machine status[67]. Alerts can be sent to their mobile devices if any issues arise, enabling a faster response. This remote visibility proved especially valuable in scenarios such as the COVID-19 pandemic, where travel was limited and on-site staffing was reduced; cloud-connected factories enabled supervision with minimal physical presence. Furthermore, multiple stakeholders (production, maintenance, supply chain, and even suppliers or customers in some cases) can collaborate using shared cloud data. For instance, a design engineer in one country can observe test results from a pilot production run in another country via cloud logs and provide feedback in near real time. Enterprise-wide visibility improves because all facilities are integrated into a single cloud system. Executives gain a comprehensive view of performance across plants, and best practices can be identified by comparing data. Additionally, remote diagnostics are enabled: if a machine malfunctions, an expert vendor or engineer can securely access its data via the cloud to help troubleshoot without the delay of travelling on-site. This connectedness enhances agility in decision-making and problem-solving. Overall, cloud and edge computing help break down geographic and departmental silos, creating a more connected and responsive organisation.

Data Analytics and Insight Generation: Manufacturing processes generate an immense volume of data, from machine sensor readings to production output counts to quality inspection measurements. Cloud computing provides the storage and computational power to harness this big data for actionable insights. By aggregating data from many sources (potentially across multiple factories), the cloud enables advanced analytics techniques, such as machine learning, statistical analysis, and simulation, to be applied. These analytics can uncover patterns and correlations that humans would miss. For example, analysing thousands of production cycles might reveal that certain environmental conditions have a slight impact on product dimensions, leading to a process adjustment for improved yield. Cloud-based analytics drive process optimisation (finding optimal operating parameters), quality improvement (identifying root causes of defects), and demand forecasting (using production and sales data to predict future needs). Edge devices contribute by preprocessing data and sending enriched, relevant information to the cloud. With the heavy lifting done by cloud analytics, manufacturers gain a deeper understanding of their operations. They can implement data-driven decision-making, where decisions on everything from factory scheduling to equipment replacement are informed by real-time data and predictive models rather than intuition alone. Moreover, with AI and machine learning models, manufacturers can achieve tasks such as computer vision for product inspection, generative design for product optimisation, and predictive supply chain logistics. These computationally intensive tasks are made feasible by cloud platforms. A telling point is that the growth of IIoT (Industrial IoT) and related technologies in manufacturing is predicated mainly on the availability of cloud services to process and make sense of the data deluge[67]. In essence, cloud computing turns raw shop-floor data into valuable intelligence, enhancing continuous improvement and innovation.

Integration with IoT and Interoperability: Cloud and edge technologies together serve as the glue that integrates diverse machines and systems in manufacturing. Modern factories often have a mix of equipment from different eras and vendors – incorporating them is a challenge. Standard protocols and IIoT gateways (usually running at the edge) can interface with legacy machines (for example, via OPC UA or MQTT protocols) and send data to the cloud, where it is unified under common data models[70]. This means even older equipment can participate in digital workflows. The result is a higher level of interoperability across the production line. Cloud platforms also often provide APIs and connectors to business systems (like ERP, PLM, CRM), enabling a smooth flow of information from the shop floor to the top floor. For instance, machine production counts can automatically update inventory levels and trigger restocking in the ERP system. The benefits of such integration include reduced manual data entry, the elimination of data silos, and real-time coordination between production and business processes. Additionally, by connecting IoT devices to a common cloud platform, manufacturers create a foundation for new capabilities, such as digital twins – virtual replicas of physical assets or processes. Data from edge devices keeps digital twin models in the cloud up-to-date, allowing simulation and troubleshooting of manufacturing processes in a virtual environment. All these integrations rely on cloud-edge architectures. When done well, manufacturers gain agility – production can quickly adjust to changes in orders or supply, because every part of the system (machines, controllers, and enterprise software) is communicating. The high visibility and synchronisation across previously disparate systems lead to benefits like better quality control (all inspection data feeds a central quality system), improved traceability for compliance, and faster new product introductions (design changes propagate quickly to production configurations via connected systems). In summary, cloud and edge computing together provide the connective tissue of smart manufacturing, bringing numerous operational benefits: less downtime, greater scalability, remote insight, data-driven optimisation, and seamless integration of the entire manufacturing ecosystem[67].

Challenges and Limitations

While cloud and edge computing offer powerful benefits, manufacturers must also navigate several challenges and limitations when implementing these technologies. Understanding these issues is crucial for successful adoption and for designing robust systems that truly meet industrial needs. Key challenges include:

Latency and Real-Time Requirements: Not all manufacturing operations can tolerate the delays inherent in cloud computing. Network latency – the time it takes for data to travel from a factory to a remote cloud server and back – can be a few milliseconds to hundreds of milliseconds, depending on the network. For many industrial control scenarios, even minor delays are unacceptable. For example, if a sensor detects a hazardous condition (like a robotic arm about to collide with an object or a person), the response needs to be near-instantaneous. Routing that signal to a cloud and waiting for a response could be too late. Manufacturers have learned that sending every control signal to the cloud introduces the risk of bottlenecks and slow reaction times, which can affect safety and quality[68]. While edge computing mitigates this by handling immediate decisions locally, it introduces complexity (requiring logic to be maintained in distributed devices). A challenge is determining which processes must run at the edge vs. what can be safely done in the cloud. Systems must be architected carefully to ensure that any control loop with strict real-time needs is kept local. Another latency-related limitation is the variability of internet connections; network congestion or outages can disrupt cloud communications. Manufacturers often invest in redundant network links or private networks (and increasingly consider technologies like 5G or Time-Sensitive Networking) to reduce latency and improve reliability, which can increase cost and complexity. The bottom line is that manufacturing demands real-time performance, and ensuring low-latency response remains a technical hurdle when the cloud is in the loop[68].

Cybersecurity Risks: Connecting factory equipment and sensitive production data to the internet or outside networks inherently increases exposure to cyber threats. Industrial systems that were once isolated can become targets for hackers or malware when they are connected to cloud services. Security is a paramount concern – a breach could not only compromise data but potentially disrupt operations or even endanger safety. Manufacturers must address both cloud security and edge device security. On the cloud side, they need confidence that the cloud provider's infrastructure is secure and compliant with industry regulations. Data travelling to and from the cloud must be encrypted to prevent interception. On the edge side, the devices themselves (sensors, gateways, etc.) can be points of entry if not properly hardened. Factories might deploy dozens or hundreds of edge devices; each needs secure access control, firmware updates, and protection against tampering. Managing this can be daunting, especially for firms new to IT security practices. Security incidents in OT (like the well-known Stuxnet attack or other industrial breaches) have raised awareness that cybersecurity in manufacturing is a serious issue once connectivity is introduced. Moreover, data ownership and privacy concerns arise: when using third-party cloud platforms, manufacturers worry about who can access their data, how it is used, and ensuring proprietary process data or trade secrets are not exposed. Some companies hesitate to put specific sensitive production recipes or quality data in a public cloud for fear of intellectual property leakage. Edge computing can alleviate some data exposure by keeping critical data on-site[69], as noted earlier, and only sending non-sensitive or aggregated data to the cloud. In highly regulated industries, data sovereignty laws might require that data stay within specific geographic boundaries, complicating the use of global cloud data centres [69]. To address security, manufacturers are adopting comprehensive cybersecurity frameworks (like the IEC 62443 standards for industrial control system security) and employing practices such as network segmentation (separating OT networks from broader IT), zero-trust security models, and continuous monitoring. Ensuring robust security end-to-end is a non-trivial challenge that comes along with cloud-edge integration.

Integration and Interoperability Issues: Many manufacturers operate "brownfield" environments – existing factories with legacy machines and systems that were never designed for connectivity. Integrating these into a cloud-edge architecture can be a complex process. Different equipment may use proprietary protocols or data formats. Even within a single plant, there may be separate systems for different functions (one for quality inspection, another for maintenance management, etc.) that historically don't talk to each other[70]. When introducing cloud and edge, companies often struggle to bring IT and OT together into a unified system[66]. Creating interfaces or middleware to translate machine data into a form usable by cloud applications (and vice versa) is a significant undertaking. Standards like OPC UA have been invaluable in addressing this challenge by providing a common communication architecture for machinery. For instance, BMW Group reported that using the OPC UA standard was key to integrating their diverse equipment and enabling interoperability across plants. Before that, each site had its own siloed systems that made data exchange difficult[70]. Even with modern protocols, deployment can be messy: some companies end up with a mishmash of hardware and software at the edge, added piecemeal as needs arose, leading to a fragmented solution that is hard to maintain[66]. Over time, without careful planning, adding more sensors and edge devices can increase complexity exponentially, making it challenging to scale further or troubleshoot problems. Essentially, infrastructure complexity and lack of standardisation can slow down cloud-edge adoption. Companies need to invest in the right platforms and potentially update legacy equipment with modern interfaces (or utilise gateway devices to bridge the gap).

This integration phase can require significant time and skilled resources, and it's a common point where projects stall.

Scalability and Management of Edge Infrastructure: While the cloud side can scale elastically, scaling out edge computing is more manual and can become a logistical headache. A manufacturer might start with a pilot of a few edge devices, but a full deployment could mean managing hundreds of edge nodes across multiple facilities. Each of these nodes may require configuration, software updates, health monitoring, and occasionally physical maintenance. Ensuring all edge devices are running the latest analytics model or that none have fallen offline requires new management tools and approaches. This is a shift for companies used to maintaining centralised servers – now the compute is distributed in many boxes on the shop floor (which may be in harsh environments, behind firewalls, etc.). Without a proper edge management platform, administrative overhead can rise sharply. Red Hat notes that gathering and handling the massive data from many edge sites is a significant challenge, as the number of sources grows, so does the strain on networks and the potential for data bottlenecks[66]. Companies also worry about future-proofing: deploying an edge solution today that might not easily expand or might become obsolete. If each production line installs a different edge solution, consolidating them later is problematic. This is why having a common, horizontal framework that spans the entire infrastructure is recommended to manage distributed data sources efficiently [66]. Solutions like centralised orchestration for edge (deploying applications from the cloud to edge nodes) are emerging to solve this, but adopting them requires an up-front commitment to a platform (which could be a challenge if companies fear vendor lock-in or have limited in-house IT expertise). In summary, operational complexity – maintaining many moving parts (devices, connections, software versions) – is a limitation that firms must plan for when extending IT into the OT domain.

Organisational and Skill Barriers: Beyond technical issues, manufacturers often face people and process challenges in implementing cloud and edge solutions. OT personnel (e.g., control engineers, technicians) may not be well-versed in IT concepts, such as cloud architecture or cybersecurity, and IT staff may not fully understand shop-floor processes and real-time constraints. Bridging this skill gap requires training or hiring for new roles such as industrial data engineers or IoT architects. There can also be cultural resistance: production teams may be wary of relying on cloud systems for their manual processes, or they may fear job displacement due to automation. Change management is crucial for gaining buy-in for these digital initiatives. Additionally, cost and ROI concerns can be a barrier – while cloud services and edge devices promise long-term savings and efficiency, there is an initial investment (in devices, networking upgrades, software subscriptions, etc.) that must be justified. If not implemented strategically, the cost of infrastructure or cloud usage fees could offset some of the benefits. Manufacturers need to calculate ROI carefully and often start with high-impact use cases to prove value (e.g., predictive maintenance usually has a clear ROI by preventing costly downtime).

In light of these challenges, many manufacturers adopt a cautious, phased approach. Starting with pilot projects, utilising hybrid architectures (retaining some on-premises capabilities while gradually leveraging cloud/edge), and involving cross-functional teams (including IT, OT, and security) to address issues holistically. Each challenge – whether it is latency, security, integration, or complexity – has solutions, but they require careful design and effective governance. For instance, latency issues can be mitigated by a well-planned division of edge vs cloud tasks; security risks can be reduced with strong encryption and access controls plus keeping the most sensitive data on-prem; integration woes ease by adopting industry standards and modernizing equipment interfaces; and managing edge infrastructure becomes easier with centralized orchestration tools (often part of a hybrid cloud platform). As manufacturing firms overcome these hurdles, they can fully capitalise on the promise of cloud and edge computing.

Real-World Examples and Case Studies

To illustrate the impact of cloud and edge computing, it is helpful to examine real-world implementations across various manufacturing sectors. Below are examples from the automotive, aerospace, and precision manufacturing industries that demonstrate the technologies in action:

Automotive Industry

The automotive sector has been at the forefront of smart manufacturing, leveraging cloud-edge solutions to improve quality and efficiency on the assembly line. A striking example comes from a leading European automotive manufacturer that implemented a comprehensive edge computing strategy in its production facilities. The deployment focused on three critical areas: welding quality, equipment maintenance, and production flow[69]. At welding stations, edge computing nodes were equipped with sensors (like thermal cameras monitoring weld temperature and power sensors measuring welding current). The edge nodes ran real-time analytics to determine optimal welding parameters and even automatically adjusted robot settings on the fly to ensure consistent weld quality under varying conditions[69]. In parallel, for predictive maintenance, other edge devices collected vibration and acoustic data from key machines (e.g., robotic arms, stamping presses) and analysed them locally to detect early signs of component fatigue or misalignment[69]. When patterns indicating potential failure emerged, maintenance teams received instant alerts with diagnostic information, enabling them to address issues before a breakdown occurred proactively. Additionally, edge systems tracked the movement of parts on the assembly line to pinpoint bottlenecks in real time; if one section was slowing down, the system could adjust conveyor speeds or reroute tasks to balance the workflow[69]. The results after 18 months were impressive: the manufacturer saw a 37% reduction in welding-related defects, a 62% decrease in unplanned downtime, and a 24% increase in overall production throughput, along with an energy consumption reduction[69]. The initial investment paid for itself in just 9 months due to the quality and uptime gains[69]. This example underscores how combining edge analytics on the factory floor with cloud oversight yields tangible improvements in an automotive plant.

Another automotive case highlights the importance of standards and the use of hybrid cloud-edge solutions. BMW, for instance, has integrated the OPC UA standard for machine-to-machine communication, along with edge computing, across its factories. They found that using OPC UA with edge gateways allowed them to pull large volumes of data (such as high-resolution camera images for inspection) and preprocess it at the edge before sending summarised information to the cloud [70]. This not only improved response times but also reduced the amount of raw data leaving the plant, addressing bandwidth and security concerns. In one BMW application, an edge computer performed particle analysis for environmental monitoring and could directly command a PLC (programmable logic controller) to adjust ventilation in real-time via OPC UA. Simultaneously, historical data were sent to the cloud for archival and deeper analysis [70]. In another example, autonomous guided vehicles communicated through an edge gateway to open/close doors as they moved through the factory, demonstrating how edge devices can promptly orchestrate physical processes [70]. These examples from the automotive world demonstrate a recurring theme: edge computing ensures rapid local control (for quality and automation), while cloud systems aggregate data and enforce standards enterprise-wide. Automakers have leveraged this combination to achieve both micro-level improvements (on individual production tasks) and macro-level coordination (across their entire production network).

Aerospace and Defence Industry

Aerospace manufacturing involves extremely complex products, lengthy supply chains, and stringent compliance requirements, making it well-suited for cloud and edge computing applications. One prominent use case in aerospace is the creation of a digital twin for production and supply chain management. For example, leading aerospace companies (like aircraft manufacturers) are shifting from linear supply chains to dynamic digital supply networks supported by cloud platforms[71]. In practice, this means data from design, manufacturing, suppliers, and even in-service aircraft are integrated. A cloud-based system compiles information on millions of parts and components, providing a virtual model of the entire production process. This virtual model is constantly updated with real-world data (e.g., sensor data from factory machines, logistics data on parts shipments) through IoT connectivity. Edge

146

devices at factories play a role by feeding real-time shop-floor data into the network. For instance, an edge device at a machine reports that a batch of components is completed, and the cloud system automatically signals the next step or updates inventory. The core advantage is agility: if a design change is made or there's a disruption (such as a delayed shipment), the system can rapidly adjust production schedules and distribution because all stakeholders are connected via the cloud [71]. During the COVID-19 pandemic, such cloud-edge ecosystems helped aerospace manufacturers cope with personnel and supply chain stresses by enabling more remote operations and reconfiguration of production lines as needed [71].

Another area is maintenance and quality assurance in aerospace manufacturing. The aerospace industry has stringent documentation and traceability needs (e.g., every part on an aircraft must be tracked). Here, edge computing can ensure data integrity and compliance. A pharmaceutical manufacturer (analogous in its regulatory rigour to aerospace) used edge computing to handle production line monitoring for compliance with Good Manufacturing Practice (GMP) requirements[69]. By processing critical quality data locally and keeping tamper-evident records at the edge, they reduced their compliance reporting workload by 40% while strengthening security. Similarly, an aerospace plant could utilise edge devices to record quality checks and equipment calibrations in real-time, and only send a secure summary to cloud storage. This creates an unbroken chain of custody for data, simplifying audits. Deloitte analysts have also described scenarios in aviation maintenance where a cloud-edge-mobile system is used: when an aircraft is at a gate, sensors (edge) might diagnose an issue, the data goes to the cloud, which then immediately notifies a technician's mobile device to take action at the airport gate[71]. The result is faster turnaround and fewer delays, illustrating how cloud-edge can extend into operational maintenance. Overall, in aerospace and defence, hybrid cloud-edge solutions help manage complex processes with high reliability. The cloud provides heavy analytics and global visibility (linking design, manufacturing, and operations). At the same time, edge computing offers the immediacy and strict control needed on the factory floor and in maintenance scenarios.

Precision Manufacturing and Small/Medium Enterprises

Precision manufacturing refers to producers of high-accuracy, often custom-engineered parts (for industries like aerospace, medical devices, automotive, etc.). These companies, even if smaller in scale, greatly benefit from cloud and edge technology to stay competitive. A case in point is Polamer Precision, a contract aerospace manufacturer producing critical engine and airframe components. Polamer adopted a cloud-based manufacturing system (the Plex Manufacturing Cloud) to run its operations and quality control. The cloud ERP/MES system provided a unified platform for managing production, inventory, quality, and the supply chain. One major driver for them was the need to handle high part variety and complexity – they have thousands of different part SKUs and hundreds of suppliers, akin to an automotive environment, but at a smaller scale[72]. By moving to a cloud platform, Polamer was able to improve visibility and coordination across all these variables. The cloud system directly interfaces with automation on the shop floor; for example, Polamer has set up automation cells that communicate with Plex in real-time, as parts are machined and inspected, data is fed into the cloud, which can then adjust schedules or update quality records immediately.[72]. The result is a highly responsive production system that minimises manual intervention. Polamer's leadership noted that the cloud architecture allowed customisation and expansion that on-premise solutions couldn't – as they grew, the system scaled, and they could tailor it to their unique processes[72]. This shows how even mid-sized manufacturers leverage cloud computing for enterprise-grade capabilities (like advanced analytics and integration) without needing a large IT department.

Edge computing is also making inroads in precision manufacturing, often via equipment monitoring solutions that can be deployed quickly. For instance, a small machine shop might install a smart edge device on each CNC machine to monitor spindle loads and surface finish quality in real time. That device could alert operators to any deviation that could affect precision, ensuring high-quality output. The aggregated data could then be sent to a cloud service that the shop owner uses to analyse productivity or share live updates with customers on job status. Such technology was historically accessible only to large factories; however, the rise of affordable IoT sensors and cloud subscription models (software-as-a-service for manufacturing) has democratised it.

In summary, across the automotive, aerospace, and precision manufacturing sectors, we observe common themes: cloud and edge computing are being deployed to enhance quality, reduce downtime, increase agility, and provide end-to-end visibility. The automotive industry emphasises production efficiency and quick feedback loops; aerospace emphasises the integration of vast, complex processes with reliability; precision manufacturers focus on leveraging these tools for agility and quality in high-mix, low-volume settings. These case studies validate the theoretical benefits discussed earlier – they show real metrics like defect rates dropping, maintenance costs shrinking, and throughput rising thanks to these technologies[69]. They also emphasise that success stems from a combination of cloud and edge, rather than relying on either in isolation.

Hybrid Cloud-Edge Strategies in Manufacturing

Rather than choosing cloud or edge, most manufacturers are finding that an optimal architecture is a hybrid model that combines the strengths of both. In a hybrid cloud-edge strategy, specific tasks and data are handled at the edge (on-premises at the factory), while others are handled in the cloud, with seamless data flow between them. The goal is to balance the trade-offs: use edge computing where ultra-fast response or local autonomy is needed, and use cloud computing for global aggregation, heavy analytics, and coordination. This hybrid approach is increasingly seen as the best of both worlds in manufacturing[66, 68].

In practice, manufacturers make decisions at multiple levels – machine-level, factory-level, and enterprise-level[68]. Edge computing is ideal for making machine-level and sometimes factory-level decisions, such as controlling a machine in real-time or running a work cell autonomously. Cloud computing excels at making enterprise-level decisions, such as comparing performance across factories, optimising supply chains, and planning for the long term. The two levels need to communicate. A robust hybrid architecture ensures that edge devices feed data to the cloud for broader analysis. Conversely, cloud insights or updates (such as a new machine learning model) are deployed back to the edge for execution.

One typical pattern is training AI models in the cloud and deploying them to the edge. A manufacturer might collect large datasets of equipment behaviour in the cloud and train predictive maintenance models or quality inspection image recognition models using the computational scale of cloud servers. Once the model is trained and tested, it is then sent down to edge devices on the factory floor, which run the model inference in real time on streaming data[69]. This way, the heavy lifting (model development) occurs in the cloud, while the real-time application takes place at the edge, achieving both accuracy (from advanced AI) and low latency. IBM used this pattern in their manufacturing plants for visual inspection systems: they managed AI models and edge devices from a central cloud location and automatically deployed updated models to hundreds of edge inferencing cameras on assembly lines[73]. This hybrid solution reduced the maintenance effort of updating devices (by 20%) and saved millions annually, while delivering AI inspection results at the edge within seconds[73]. IBM's example underlines how cloud management, combined with edge execution, can dramatically improve scalability – thousands of edge devices can be updated via a single cloud interface – and performance, as each device operates with minimal latency for its local task.

Another aspect of hybrid strategies is determining which data should remain at the edge and which data should be sent to the cloud. Manufacturers often employ a tiered approach, where critical data is processed at the edge, and only summaries or exceptions are sent to the cloud. This reduces bandwidth usage and costs, while also addressing concerns about data privacy. For instance, high-frequency sensor data might be too voluminous to send constantly, so the edge computes aggregate metrics over a shift and sends those. However, if an anomaly is detected, the raw snippet surrounding that event may be sent to the cloud for further analysis by engineers. Manufacturers might also choose to keep specific proprietary process data on a private edge or on-premise server (sometimes called a private cloud if it's a local data centre) and use a public cloud for less sensitive information or cross-company functions. This forms a hybrid cloud environment – mixing on-premises private clouds with public clouds – linked by the edge computing layer[66]. For example, a company could use a private cloud on-site to store detailed production data and run immediate analytics, but periodically push summary data to a public cloud where machine learning algorithms refine predictive models using data from all plants.

Trade-offs are continuously managed in such hybrid models. A key trade-off is latency vs. centralisation: moving more tasks to the edge lowers latency but increases local management; moving more tasks to the cloud simplifies central control but can introduce latency. The hybrid mindset is to allocate each function to the optimal place. A rule of thumb emerging in the industry is: "Compute as close as necessary, and as centralised as possible." That means if something can tolerate a bit more delay and benefits from aggregation, put it in the cloud; if something needs instantaneous reaction or must remain local (for security or reliability), keep it at the edge. Many manufacturers use edge for immediate control decisions but rely on the cloud for supervisory control and analytics. For example, in-cycle decisions (instant machine adjustments) are made at the edge, whereas end-of-cycle or shift-level decisions (like adjusting the day's production plan) can be made in the cloud[68].

Emerging platforms facilitate the integration of cloud and edge. Major providers (AWS, Azure, Google) offer IoT and edge services that tie into their clouds – for instance, AWS Greengrass or Azure IoT Edge allow cloud logic to be containerised and run on edge devices, syncing with the cloud when needed. Likewise, industrial automation companies offer systems to bridge these worlds. The concept of "open hybrid cloud" is advocated by companies like Red Hat, where open source platforms (e.g., Kubernetes and containerization technologies) are used to deploy applications uniformly to cloud or edge[66]. This avoids the problem of having completely separate software stacks and allows scaling applications from the cloud down to tiny edge devices with minimal changes. It also helps tackle the earlier-mentioned complexity challenge by providing a common way to manage infrastructure and applications across all levels. Red Hat notes that edge computing becomes more powerful when used in concert with hybrid cloud infrastructure – it provides flexible compute resources and unified management, making it easier to adapt and scale as needs evolve[66].

From an organisational perspective, hybrid models also help address risk: manufacturers can keep critical operations running on-premises (for safety and uptime), which appeases OT concerns, while still leveraging cloud innovation for strategic gains, which satisfies IT and business units. Many early reservations about the cloud in manufacturing (latency, security) are mitigated by keeping a foot on the ground (edge). At the same time, edge-only strategies are limited to local scope, so linking to the cloud maximises value by gleaning broader insights and enabling remote collaboration. Therefore, leading manufacturers now view cloud and edge not as an either/or decision, but as complementary components of their digital strategy[68]. Industry analysts predict that the most successful manufacturers will adopt integrated approaches where edge computing is part of a comprehensive strategy rather than an isolated initiative[69]. Indeed, as one source put it, the question is no longer whether to use edge computing, but how quickly you can implement it alongside the cloud to gain a competitive edge[69].

In summary, hybrid cloud-edge models enable manufacturers to maximise the potential of both approaches while mitigating their respective limitations [68]. Properly designed, a hybrid architecture provides agility – the proper processing at the right place – and resilience, allowing the system to continue operating under various conditions. As we move forward, such architectures are expected to become the norm in industrial deployments, supported by new tools and standards that make the cloud-edge boundary increasingly seamless.

Standards and Future Trends

The landscape of cloud and edge computing in manufacturing is continuously evolving. Industry standards, emerging technologies, and trends are shaping how these systems are implemented and the capabilities they offer. This section examines some of the key standards and consortia that influence interoperability, as well as emerging trends that indicate the direction the future is headed.

Interoperability Standards and Frameworks: Manufacturing enterprises rely on standards to ensure that machines, devices, and software from different vendors can work together in a cloud-edge architecture. One of the most significant standards is OPC UA (Open Platform Communications Unified Architecture). OPC UA is widely adopted for machine-to-machine communication in factories, providing a common language for industrial equipment. It has been instrumental in bridging the gap between OT and IT systems. Companies like BMW have reported substantial improvements in integrating disparate systems using OPC UA into a unified framework.[70]. Importantly, OPC UA includes security features (encryption, authentication) that are valuable when implementing cloud and edge solutions, as they help protect data in transit[70]. In cloud-edge contexts, OPC UA often runs on edge gateways to interface with legacy equipment and then transmits data to the cloud in a structured manner. The OPC Foundation and industry working groups also develop companion specifications (for robots, CNC machines, etc.) to enable the use of standard data models, thereby accelerating integration. Another relevant set of standards is messaging protocols, such as MQTT (a lightweight publish/subscribe protocol), which is used by many IIoT devices to send data to cloud brokers efficiently. MQTT's design for low bandwidth and reliable delivery makes it ideal for edge devices pushing sensor updates to the cloud.

To structure how cloud and edge fit into broader enterprise architecture, organisations like the Industrial Internet Consortium (IIC) have published reference architectures and frameworks (e.g., the IIC's Industrial Internet Reference Architecture), emphasising a layered approach with edge nodes feeding into cloud systems in an Industrial Internet of Things (IIoT) environment[69]. The IIC's guidance on "distributed computing in industrial systems" helps companies plan what functionality belongs at each layer (edge, platform, enterprise). Similarly, the German-led Industry 4.0 initiative and its Reference Architectural Model for Industry 4.0 (RAMI 4.0) outline a three-dimensional layer model that includes an

"Information layer" and "Functional layer" mapping roughly to cloud services, and a "Physical layer" corresponding to shop-floor devices (edge), ensuring that standards and interfaces connect these layers. Open Manufacturing Platform (OMP), a consortium including BMW, Microsoft, and others[70]. Working on open standards for IoT connectivity and semantic data models to facilitate the integration of cloud and edge computing in manufacturing. These efforts all aim at interoperability – so that, for example, an edge sensor from one vendor can communicate through an OPC UA server and be understood by a cloud application from another vendor, with minimal custom adaptation.

On the cybersecurity front, standards like IEC 62443 provide guidelines for securing industrial automation and control systems, which is highly pertinent when those systems interface with cloud/edge. Compliance with such standards means building security in every layer (device security, network security, application security) – manufacturers moving to cloud-edge must ensure their vendors and implementations meet these norms for safe operation.

In summary, the presence of robust standards (e.g., OPC UA, MQTT) and reference frameworks is making it easier to deploy cloud and edge solutions that are interoperable and secure, thereby preventing vendor lock-in and allowing for scalable integration. Manufacturers are advised to favour solutions that adhere to these open standards to future-proof their investments and enable multi-vendor ecosystems.

Emerging Technologies and Trends

Looking to the future, several trends are poised to transform cloud and edge computing in manufacturing further:

Edge-to-Cloud Continuum and Hybrid Architectures: The strict line between edge and cloud is blurring with the development of technologies that allow more fluid distribution of workloads. We will see increasingly seamless edge-cloud hybrid architectures, where data flows effortlessly from edge devices to cloud and back, and orchestration systems manage this distribution dynamically[69]. For example, as 5G networks roll out in factories (more on this below), certain computations might run wherever it's most efficient at the moment – sometimes at a local micro data centre (a concept often termed "fog computing"), other times in a regional cloud – all abstracted from the end user. The trend is towards unified management: the operator doesn't have to manage edge and cloud separately, but uses one platform to deploy services across both. This will significantly simplify the expansion of digital initiatives.

5G and Advanced Networking: The adoption of private 5G networks in manufacturing plants is a game-changer for connectivity. 5G promises ultra-reliable low-latency communication (URLLC), which can enable wireless connections that are as fast and dependable as wired ones. This is ideal for edge computing because it allows more devices to communicate with minimal lag. With 5G, an edge device on a moving robot or AGV (automated guided vehicle) can send high-bandwidth data (like video) to a nearby server or cloud with only a few milliseconds of delay, meeting real-time needs. Private 5G networks (spectrum dedicated to the enterprise in the factory) are emerging, enabling manufacturers to have complete control over their network performance. Combined with technologies like Time-Sensitive Networking (TSN), which brings deterministic timing to standard Ethernet, the factory network of the future will support synchronising edge devices and the cloud for time-critical tasks[74]. This networking leap will likely expand the use of edge computing (more mobile and distributed devices can be connected reliably) and also allow more data to be sent to the cloud for advanced analysis without congestion. Essentially, 5G will act as an enabler that ties cloud and edge computing closer together, handling the massive data flows of tomorrow's smart factories with ease [69].

Artificial Intelligence and Machine Learning at the Edge: AI is increasingly being deployed at the edge in manufacturing. This includes computer vision systems for inspection, AI controllers for robotics, and anomaly detection algorithms for equipment, running directly on edge hardware (sometimes specialised AI chips). The trend is that AI models are trained on cloud infrastructure using big data, and then pushed out to edge devices for real-time execution[69]. This trend will accelerate as tools for automated machine learning and model optimisation continue to improve. We may see more TinyML (Tiny Machine Learning), where even microcontrollers on devices can run simple models. The benefit is more intelligent local behaviour: machines that can self-optimise and adapt on the fly. For example, an intelligent edge sensor might run an AI model to decide if a part is within tolerance, without needing to send raw measurements to the cloud each time. As AI becomes more prevalent, edge devices will need to handle more complex processing (leading to more powerful industrial edge hardware), and cloud systems will play a role in centrally managing these distributed AI instances (updating models, aggregating learning, e.g., federated learning across multiple machines). AI at the edge will also feed into augmented reality (AR) and assisted operations – think of an engineer wearing AR glasses in a factory, where the glasses' edge processor recognises equipment and overlays data. At the same time, the cloud provides the latest analytics for that equipment. This convergence of AI, edge, and cloud will make factories more autonomous and empower workers with more information.

Digital Twins and Simulation: The concept of digital twins – virtual models of physical assets – is gaining traction in manufacturing. Cloud computing provides the environment to host detailed digital twins of machines, production lines, or entire facilities, because it requires extensive computation and data integration. Edge computing provides the live data that keeps the twin updated in real time. The trend is towards integrating digital twins with edge data for real-time mirroring of operations[69]. For instance, a digital twin of a CNC machine might run in the cloud, continuously fed by edge data from the actual machine. Operators can use the twin to simulate adjustments or predict outcomes (such as how a machine would behave if a parameter were changed) and then apply the best settings in the real world. In the future, almost every critical piece of equipment might have a cloud-based twin fed by edge devices, enabling predictive diagnostics and what-if simulations on demand. This will improve decision-making and allow virtual commissioning of changes (testing them digitally before physical implementation).

Autonomous and Self-Optimising Systems: As edge and cloud capabilities mature, we will edge closer to fully autonomous manufacturing systems. Already, we see elements like autonomous material handling (AGVs and robots coordinating via edge controllers) and self-optimising process lines (that adjust settings in real time). The emerging trend is edge-powered autonomy, where local control loops can handle complex tasks without human input, yet remain supervised by cloud AI that provides high-level guidance[69]. In the next few years, manufacturers aim for production lines that can reconfigure themselves for different products with minimal intervention, using edge computing to control robotics and equipment adjustments, and cloud-hosted intelligence to schedule and orchestrate the changes. This ties into Industry 4.0's vision of highly flexible "batch size 1" manufacturing. Cloud and edge will serve as the nervous system of these agile factories: edge nodes acting like reflexes and muscle movements, cloud systems like the brain aggregating information and sending directives.

Enhanced Security and Distributed Cloud: Finally, in response to security concerns and the need for low latency, we may see more distributed cloud infrastructure, where cloud providers offer mini data centres at the edge (sometimes called cloud edge or MEC, multi-access edge computing). These could be on-premises or at local telecom exchanges, bringing cloud services physically closer to factories. This can reduce latency and give companies more control over data location. Paired with advanced encryption, blockchain for data integrity, and other emerging security tech, the future cloud-edge environment will be more secure and trusted. One trend is the use of blockchain in manufacturing networks for secure, verifiable data sharing among partners (though not directly a cloud-edge technology, blockchain can be hosted in the cloud and record data from edge devices, providing tamper-proof logs, especially useful in supply chain or quality certifications).

In essence, the future points to a convergence of technologies: faster networks (5G, TSN), more intelligent algorithms (AI, digital twins), and more distributed computing (edge and cloud continuum). Standards and consortia will continue to play a key role in ensuring these pieces fit together. Manufacturers can anticipate even greater efficiency and capabilities as these trends mature. The factories of tomorrow are envisioned to have intelligence everywhere – from sensors with built-in AI to cloud systems that coordinate entire production ecosystems, operating in harmony.

For industry professionals, staying abreast of these trends is vital. It means investing in systems that are flexible and upgradable. Many are already updating their infrastructure to be ready: adopting containerization and microservices so that the same application can run on cloud or edge, deploying private 5G testbeds in their plants, training staff in data science and AI for manufacturing, and participating in industry groups to shape standards. As one report observed, edge computing is likely to become an essential foundation, rather than an optional add-on, for future manufacturing excellence[69]. The trajectory is clear – cloud and edge computing will continue to drive innovation in manufacturing, enabling factories to be more intelligent, responsive, and integrated than ever before.

Digital Twin: The Virtual Mirror of Manufacturing

A Digital Twin is a virtual representation of a real-world entity, such as a physical asset, system, or process, that is continually updated with data to mirror its current condition and behaviour. In manufacturing, this means creating a highly detailed digital replica of shop-floor elements (machines, production lines, products, etc.) that stays in sync with its physical counterpart in real time[75, 76]. Unlike static models, a digital twin is a "living" model: it receives input from sensors and operational data, enabling it to reflect real-world changes in operating conditions, environment, and performance[75]. Through this constant data flow between the physical and digital realms, the digital twin evolves alongside the physical object, spanning the entire lifecycle from design and production to operation and maintenance[77, 78].

In practical terms, a digital twin in manufacturing can be a virtual replica of a robotic assembly cell, a CNC machine, an entire production line, or even an entire factory. It integrates real-time sensor data, historical data, and physics-based models to provide an accurate, dynamic mirror of the state of the physical system[76]. For example, Airbus defines a digital twin as "more than just a digital model; it's a dynamic, living virtual replica of a physical object, process, or system" that integrates data from design, production, and in-service operations to give a continuous real-time reflection of its real-world counterpart[77]. This dynamic nature allows the twin to simulate the physical system's response to various inputs and conditions.

Key characteristics of digital twins include, (1) a two-way data connection with the physical asset (incoming data to update the twin, and sometimes control or insights fed back to the physical world), (2) use of advanced simulations and analytics (often leveraging AI/ML algorithms) to model behavior and predict outcomes, and (3) a virtualization environment that can be interacted with by users or other digital systems[75, 78]. It's important to distinguish a digital twin from a simple simulation model. Traditional simulations are typically run in isolation for a specific scenario; a digital twin, by contrast, continuously synchronises with live data and can run myriad simulations in parallel, providing richer insights and real-time decision support[78]. As McKinsey puts it, a digital twin "updates in real time to reflect the original version" and can be used to create an immersive virtual environment linking every aspect of an organisation's operations for better decision-making[75].

In summary, in manufacturing contexts, a digital twin serves as a data-driven virtual counterpart of factory assets and processes. It enables engineers and managers to monitor status, run virtual experiments, and obtain predictive insights without disrupting the physical production. This concept, foundational to Industry 4.0 and smart manufacturing, is increasingly seen as critical for bridging physical and digital systems, allowing manufacturers to build products and optimise operations in the digital world before (and while) doing so in the real world.

Integration of Digital Twins with Manufacturing Processes and Systems

Implementing digital twins in manufacturing requires integration across multiple layers of technology and processes. At its core, it involves integrating shop-floor data sources, analytical models, and enterprise systems into a cohesive framework. A typical digital twin architecture for manufacturing can be viewed in layered terms[76]:

Data Layer (Physical Layer): This includes the physical observable manufacturing elements (equipment, tools, work cells, etc.) outfitted with IoT sensors and data acquisition systems. These devices collect real-time data (e.g., temperatures, pressures, speeds, positions, etc.) and feed it into the digital twin [76]. Integration at this layer involves retrofitting machines with sensors and connecting them to existing Industrial IoT infrastructure, making streams of shop-floor data (from PLCs, SCADA systems, MES sensors, etc.) accessible. In a standards-based framework like ISO 23247 (Digital Twin Framework for Manufacturing), this corresponds to "observable manufacturing elements" and device communication interfaces that gather state data from the shop.

Model Layer (Virtual Twin Layer): This is where the core digital twin models live. The incoming raw data is aggregated and processed (often via an edge or cloud platform) into the virtual representation of each asset or processap238.org. The model layer uses physics-based simulation models (e.g. finite element models of machines, kinematic models of robots, or process simulation models) combined with data-driven models (AI/ML algorithms that learn from historical data) to update the twin's state and behaviour in real time[76]. Here, integration involves linking the twin to engineering design data and analytical tools. For instance, CAD/CAE models and manufacturing process models feed into the twin's baseline, and are continuously refined with live data. The twin might use machine learning models to predict outcomes (like quality or failures) based on the sensor inputs. This layer often interacts with a cloud computing environment for heavy computations and data storage, while critical real-time updates might be handled on the edge (near the equipment) for low latency[79, 80]. In practice, manufacturers integrate their MES (Manufacturing Execution Systems) and control systems with the twin so that the virtual model gets triggers from production events, and can also send control adjustments. For example, a CNC machine's digital twin may be connected to the machine's CNC controller via an IIoT gateway, enabling real-time monitoring and even closed-loop control adjustments based on twin simulations.

Service Layer (User/Application Layer): At the top, the insights from the digital twin are delivered to users and enterprise applications[76]. This includes dashboards for real-time monitoring, analytical applications for performance analysis, and integration with enterprise software like PLM (Product Lifecycle Management), ERP (Enterprise Resource Planning), and SCM (Supply Chain Management) systems. By integrating the twin here, organisations create a "digital thread" through the product lifecycle – linking design data (PLM) with manufacturing data (MES/SCADA) and operational data (field or customer usage) via the twin[77]. For instance, design engineers can use the twin to test design changes and immediately see the impact on production; meanwhile, production planners can use the twin to simulate schedule adjustments given supply chain data. The twins' outputs (e.g., predicted maintenance needs, quality alerts, optimal process parameters) can be automatically fed into maintenance management systems or quality control workflows. Modern manufacturing digital twins often integrate with cloud-based IoT platforms (such as Siemens MindSphere, GE Predix, or Airbus's Skywise), which act as hubs for data and provide APIs for various business systems[77].

Integration is not only technical, but also process integration. Digital twins must be woven into the operational routines of the factory. For example, before implementing a process change on the line, engineers might run scenarios on the digital twin first (making the twin part of the standard operating procedure for engineering changes). Or maintenance crews may rely on the twins' predictive analytics output as part of their daily maintenance planning. In essence, the digital twin becomes an integral part of both the control loop on the shop floor (feeding data to and from controllers for real-time adjustments) and the decision loop at the management level (informing human decisions in production planning, scheduling, and design). Achieving this integration can be complex – it often requires interfacing the twin with legacy systems (older machines and databases), which may involve custom adapters or the use of standards like OPC UA for exchanging machine data. It also benefits from adopting reference architectures: for example, ISO 23247 provides a reference architecture for manufacturing twins to guide how devices, twin models, and user applications interactap238.orgap238.org. Companies like Siemens and NVIDIA have demonstrated "factory digital twin" reference architectures that show data flowing from the factory floor to a simulation environment (e.g., NVIDIA's Omniverse for factory simulation) and back to operators, illustrating the practical integration of these layers.

In summary, integrating a digital twin into manufacturing means connecting operational technology (OT) with information technology (IT). It links sensors and control systems on the factory floor with big data and analytics platforms, and further with business applications. When done successfully, the digital twin serves as the centrepiece of a smart manufacturing ecosystem, ensuring that information flows seamlessly from physical operations to digital analytics and back again, thereby creating a closed feedback loop for continuous improvement.

Benefits of Digital Twins in Manufacturing

Digital twins are transformative because they enable manufacturers to understand and optimise their processes in ways that were previously impossible. Key benefits include improved process optimisation, predictive maintenance, enhanced quality control, and holistic lifecycle management, among other benefits. These benefits translate into tangible outcomes, including higher productivity, reduced downtime, lower costs, and improved product quality. Below, we discuss each significant benefit in detail:

Real-Time Monitoring and Process Optimisation: A digital twin provides real-time visibility into manufacturing operations, allowing for on-the-fly optimisation. By mirroring the live status of machines and production lines, the twin enables continuous monitoring and control. Operators and control algorithms can utilise the twin to detect deviations or inefficiencies and adjust parameters accordingly immediately. For example, in a CNC milling operation, a digital twin might continuously collect data on cutting forces, tool temperature, and tool wear, and then dynamically adjust cutting parameters (such as feed rate and spindle speed) to optimise machining efficiency and maintain part quality [81]. If the twins' analytics detect excessive tool wear, it can even trigger a tool change or machining adjustment before a quality issue arises, thus ensuring uninterrupted production [81]. This capability to simulate and fine-tune processes virtually leads to better resource utilisation and throughput. Manufacturers can test "what-if" scenarios on the twin (such as reconfiguring a production line or changing process parameters) without disrupting the real line, identifying the most efficient configurations. One case study in an automotive production line demonstrated a 6% increase in line efficiency after implementing optimisations informed by a digital twin model[82]. Similarly, manufacturers using twins for process optimisation have reported significant gains in output; for instance, McKinsey noted some companies achieved 15–20% increases in production throughput by systematically using digital twins to de-bottleneck processes[83] (a figure echoing industry case reports). In complex assembly operations (like aerospace), digital twins of the assembly line allow engineers to simulate changes in line layout or workflow digitally, resulting in more agile and flexible manufacturing systems that can be rebalanced or reconfigured much faster to meet changing demands[77].

Predictive Maintenance and Asset Health Management: One of the most celebrated benefits of digital twins is their ability to enable predictive maintenance. By continuously analysing sensor data and equipment conditions in the twin, manufacturers can predict failures before they happen and schedule maintenance proactively[82]. The twin acts as a digital condition-monitoring system, often augmented with AI algorithms that recognise patterns leading to failures. For example, a digital twin of a machine tool can ingest vibration and temperature data from the machine's bearings and use an ML model to predict when a spindle bearing is likely to fail[82]. Maintenance can then be performed during planned downtime, thus minimising unplanned outages. This predictive capability reduces maintenance costs (by preventing catastrophic failures and extending equipment life) and improves overall equipment effectiveness. In practice, companies adopting twin-driven predictive maintenance have seen dramatic results – downtime reduction is a common metric. A real case in an automotive plant showed an 87.5% reduction in downtime after implementing a digital twin for machine health monitoring[82]. Similarly, Airbus reports that in their factories (e.g., at the Saint-Eloi plant in Toulouse), machine digital twins help detect anomalies in drilling and milling operations early, allowing maintenance to be scheduled before breakdowns occur[77]. Moreover, once products are deployed, their digital twins can continue to support maintenance in the field: for instance, over 12,000 Airbus aircraft are connected via digital twin technology on the Skywise platform, enabling airlines to predict component wear and optimise maintenance schedules, thereby reducing aircraft downtime and extending component life[77]. This lifecycle approach to maintenance –

tracking an asset's health from factory to field with a twin – ensures reliability and performance for end customers as well.

Quality Control and Defect Reduction: Digital twins significantly enhance quality assurance in manufacturing processes. By simulating production and comparing expected versus actual results in real-time, a twin can identify quality deviations early. For example, sensors on a production line might feed dimensional measurements and images of a product into its digital twin. The twin can run a virtual inspection by comparing this data against the product's design specifications or an ideal digital model[81]. If any parameter – say a hole diameter or surface finish – exceeds tolerance, the system can alert quality engineers or automatically adjust the process. In the CNC machining scenario, the twin measuring tool positions and forces can detect if a part is drifting from specifications and suggest tool offset adjustments or flag the part for inspection [81]. Early detection of defects in a virtual environment prevents large batches of scrap and reduces rework, thereby saving both cost and time. Furthermore, digital twins allow extensive virtual testing of product designs and manufacturing steps before physical trials. In automotive manufacturing, for instance, automakers use digital twins to run virtual crash tests and assembly simulations. Ford has cut its vehicle design time by 25% by using digital twin simulations in place of some physical prototypes [84]. By the time a process is run in reality, many potential quality issues have been ironed out in the twin. General Motors also leveraged digital twin predictive analytics to improve quality control by 15%, catching issues earlier in the production process [84]. The twin thus acts as an ever-vigilant quality inspector and process optimiser, contributing to consistent product quality and compliance with standards.

Faster Product Development and Lifecycle Management: Digital twins integrate with the product lifecycle from design to deployment, supporting what's often called digital thread continuity. During product design and development, a digital twin of the product (or of the production system that will build it) allows engineers to verify designs and manufacturability virtually. For example, aerospace companies simulate the behaviour of an aircraft design under various conditions using a digital twin, thereby reducing the need for multiple physical prototypes [77]. Airbus credits its use of digital twins in early design and industrialisation phases with accelerating time-to-market and reducing design errors, as seen in their A320 family, where a fully digital approach (using 3D master models and twins) significantly shortened design and production lead times while cutting quality issues[77]. In automotive, BMW improved production planning efficiency by 30% using virtual simulations (a type of digital twin of the production system) before physical line changes[84]. This tight feedback between design and manufacturing, facilitated by the twin, ensures that when a product enters actual production, both the product and the process have been optimised together in the virtual domain. Throughout the product's operational life, its digital twin (especially for complex equipment like aircraft engines, turbines, or cars) continues to gather performance data. This helps manufacturers and end-users manage the asset's lifecycle – scheduling upgrades, improving future designs based on real-world performance, and even supporting end-of-life decisions (such as remanufacturing or recycling), guided by data from the twin. In essence, digital twins support Lifecycle Management by providing a unified data model of an asset from its inception (design) to its retirement, ensuring that information and learnings carry through each stage. This reduces information silos and duplication; for instance, the exact digital twin that informs manufacturing can later inform service and then provide feedback to inform the design of the next-generation product [77].

Supply Chain and Operational Planning: Beyond the factory floor, digital twins can be scaled up to model entire production networks and supply chains. By linking together twins of machines, production lines, inventory systems, and logistics, companies can simulate end-to-end operations. A supply chain digital twin can integrate real-time data on raw material availability, production rates, and shipment logistics to optimise production planning and inventory management[81]. For example, before ramping up production for a new product, a manufacturer can use a supply chain twin to identify potential bottlenecks in supplier deliveries or transportation and proactively mitigate them. This leads to reduced lead times and more resilient operations. During disruptions (like a sudden supplier outage), the twin can be used to run scenarios for re-routing orders or adjusting production schedules, thus improving response agility. Some manufacturers have begun to recognise the benefits of these holistic twins. By modelling supply and production together, they can achieve leaner inventories and shorter cycle times while avoiding stockouts, ultimately improving on-time delivery performance.

Energy Efficiency and Sustainability: Digital twins are also being applied to optimise energy usage and support sustainability goals. A twin can monitor the energy consumption of equipment and processes in real-time, identifying opportunities to reduce waste [81]. For instance, in a process involving heavy machinery, the twin can track electricity use, heating/cooling loads, and other resource inputs. By simulating adjustments (such as turning off idle machines, adjusting process timing to off-peak hours, or optimising motor loads), it can recommend strategies to reduce energy consumption while maintaining output [81]. Smart factories use these insights to cut costs and lower their carbon footprint. This has become a notable trend: according to industry analyses, many manufacturers are deploying digital twins specifically to meet sustainability targets, such as minimising energy per unit produced and monitoring carbon emissions in production[85]. By providing a detailed accounting of resource usage and enabling virtual testing of greener process alternatives, digital twins help embed sustainability into operational decisions.

Overall, the benefits of digital twins in manufacturing can be summarised as greater agility, efficiency, and insight across the production lifecycle. Companies report improvements like 20% decreases in production downtime and significant cost savings from optimised maintenance and fewer quality issues[84]. Importantly, digital twins also foster better cross-functional collaboration. Since the twin serves as a single source of truth about a product or process (in a digital format that anyone can access), it breaks down silos. Design engineers, process engineers, quality teams, and maintenance crews can all collaborate through the digital twin, trying out changes and sharing insights in the virtual space, which ultimately leads to better decisions and innovation. Airbus highlighted this collaboration benefit – their teams use shared digital twins so that, for example, designers and shop-floor workers can virtually review changes together, leading to faster and more confident implementation of improvements[77]. In short, digital twins drive continuous improvement by enabling a cycle of monitoring, analysis, and optimisation that continually loops, yielding compounding gains in productivity and quality for manufacturers.

Major Challenges and Limitations

Despite the compelling benefits, implementing digital twins in manufacturing presents challenges. The technology is complex, and many organisations encounter obstacles related to technical integration, data management, skills, and cost. Key challenges and limitations include:

Implementation Complexity & System Integration: Developing and integrating a digital twin into existing manufacturing systems can be highly complex. Manufacturing environments often have heterogeneous equipment and legacy IT systems. Ensuring a twin can interface with all relevant systems – from old PLCs on machines to modern cloud analytics – requires overcoming compatibility issues. Many early adopters find that achieving interoperability between different data sources and software is a significant hurdle[81]. Custom middleware or the adoption of standards is often necessary to connect IoT sensors, databases, simulation software, MES, ERP, and other systems into a cohesive twin. The complexity of modelling an entire manufacturing system with all its moving parts is also non-trivial; it may require capturing multi-disciplinary aspects (mechanical, electrical, control logic, etc.). High system complexity translates to high development effort and cost. In some cases, retrofitting older machinery with sensors or digital interfaces is expensive or technically challenging. This complexity and integration work can make initial implementations costly and time-consuming[81]. For example, creating a unified twin of a complete production line may involve integrating dozens of machines, each speaking a different protocol. Without careful planning, the project can become overwhelming.

Data Integration and Quality: Digital twins are only as good as the data they ingest. Ensuring data quality, consistency, and availability is a significant challenge[81]. Manufacturing data can come from various sources (sensors, manual inputs, MES logs, etc.), often in different formats and frequencies. Combining these into a coherent, real-time picture can be challenging. Problems such as missing data, noisy sensor signals, or inaccurate measurements can lead to an unreliable twin. Maintaining data integrity (so that the twin's state truly reflects reality) requires robust data validation and cleansing pipelines. Moreover, historical data is needed to train predictive models and validate simulations, and gathering sufficient historical data can be an obstacle for new twin deployments. Many manufacturers also struggle with data silos – data might exist in separate systems (maintenance logs in one database, quality data in another). Breaking down these silos to feed the twin involves organisational effort. Compatibility of data formats is a related issue; the lack of standardised data models means that integration may require a significant amount of custom translation. The Asset Administration Shell (AAS) concept from Industry 4.0, along with standards such as OPC UA information models, aims to standardise the structure of asset data for easier digital twinning. However, the adoption of these standards is still in progress. Without high-quality, well-integrated data, the digital twin's insights and predictions won't be trustworthy, limiting its usefulness.

Data Security and Privacy: Because digital twins aggregate sensitive operational data and often connect to critical control systems, they introduce security concerns[81]. A twin for a manufacturing line may contain detailed information on how that line operates (which could be IP-sensitive) and may allow control commands to be sent back to the machines. This makes it a target for cyberattacks. Manufacturers must ensure that access to the twin (and its data streams) is secure – authentication, encryption, and network security are all essential. The risk of a breach could mean not only stolen data but also the potential manipulation of the twin or even the physical process. A compromised twin could feed false information to operators or send erroneous control actions. Therefore, robust cybersecurity measures are necessary, which adds to the implementation complexity. Additionally, suppose the twin extends to the supply chain or product in the field use. In that case, there may be privacy considerations (for instance, sharing operational data with third-party service providers requires trust and possibly compliance with data regulations). Companies have to invest in secure IoT architectures and perhaps adopt emerging "digital twin security" standards that outline best practices for protecting twin environments[81]. This is a developing area, as noted by NIST: security frameworks and standards are being considered to ensure digital twin deployments are resilient to cyber threats[81].

Scalability and Performance: As a digital twin grows (with more devices, more detail, and more analytics), ensuring it remains performant and scalable is challenging [81]. A twin for a single machine might be straightforward, but what about a twin for an entire factory with thousands of sensors? The volume of real-time data can be enormous, requiring a scalable data processing infrastructure (often cloud-based). Ensuring low-latency updates for time-critical twins (e.g., those used for real-time control) may require deploying edge computing resources and optimising networks [79, 80]. Additionally, as more functionality is added to a twin (for example, adding a supply chain model on top of a factory model), the architecture must scale without becoming unstable or overly complicated. Poorly scalable designs may work as pilot projects but falter when scaled up to full production. Manufacturers often need to choose which aspects to model at high fidelity and which can be simplified to keep the twin manageable. The computational load of running detailed physics simulations and AI models continuously is non-trivial; it may require HPC (High-Performance Computing) resources or efficient model approximation techniques. Achieving real-time (or near real-time) performance as the system grows is an ongoing technical challenge.

Interoperability and Lack of Standardisation: The manufacturing world currently lacks universally adopted standards for digital twin data and interfaces. This interoperability gap means that a twin built on one platform may not easily integrate with another, or components from different vendors may not "talk" to each other natively[81]. A lack of standards can lead to vendor lock-in and slow down multi-vendor implementation projects. For example, if one machine's twin model is in format A and another's in format B, combining them into a higher-level twin could require significant effort. Standards organisations have recognised this and are working on frameworks (e.g., the ISO 23247 series, specifically for manufacturing, which defines a reference architecture and common terminology [81]). The Digital Twin Consortium (an industry consortium) is also driving efforts to harmonise definitions and interfaces, defining a digital twin as "a virtual representation of real-world entities and processes, synchronised at a specified frequency and fidelity" and promoting interoperability across industries[86]. Until such standards mature and gain widespread adoption, companies implementing digital twins often have to develop custom integration code or use middleware to bridge the gaps, which adds to the cost and complexity of the process. A related challenge is legacy systems – older equipment might not support modern protocols needed for integration, and without a standard approach, each case becomes a one-off engineering project.

Model Validation and Trustworthiness: A digital twin is only helpful if it accurately represents the physical system it is intended to replicate. Ensuring the twins' fidelity, through verification and validation (V&V), is a non-trivial challenge[81]. If the twins' simulations or predictions are wrong, it could lead to bad decisions (for example, a false prediction of failure or an incorrect process adjustment). Manufacturers need to validate that the twins' behaviour matches real-world outcomes under various conditions. This often requires extensive testing and continuous calibration of the twins' models. As conditions change (tools wear down, machines age, new product variants are introduced), the twin must be updated to remain an accurate mirror. Trust in the digital twin is essential for users to rely on it; building that trust may require a phased approach, where the twin's recommendations are cross-checked against real-world results over time. The ASME (American Society of Mechanical Engineers) has recognised this and is working on guidelines for V&V and uncertainty quantification for digital twins in manufacturing[87]. Aiming to help practitioners assess how much confidence to place in twin predictions. Without proper V&V, there's a risk that a digital twin could deviate from reality over time (model drift), especially if underlying processes change and the model isn't adjusted. Establishing a rigorous V&V procedure and possibly utilising standards (such as ASME's V&V40 for computational models, adapted to digital twins) can mitigate this. Still, it adds another layer of effort to the twin development lifecycle.

Skilled Workforce and Cultural Change: Implementing and utilising digital twins requires a multidisciplinary skill set – data scientists, simulation engineers, IoT specialists, and domain experts must all collaborate. Many manufacturers face a skills gap, lacking in-house expertise to develop advanced simulations or machine learning models for the twin. Hiring or training staff, or bringing in consultants, increases costs. Moreover, adopting a digital twin often necessitates cultural change within the organisation. Traditional engineers and operators might be accustomed to making decisions based on experience and simpler tools. Trusting a digital model's outputs or changing workflows to incorporate the twin (e.g., consulting the twin for every process change) can meet resistance. It takes time and proof of value to get buy-in from all stakeholders, from the shop floor technicians to upper management. Organisations may need to start with pilot projects and champions who advocate for the technology, gradually integrating it into standard operating procedures. The cross-functional nature of twins also means breaking silos between IT and OT departments, which can be a cultural hurdle in itself. Interdisciplinary collaboration is required (IT, operations, engineering, quality, etc., all working together via the twin), and coordinating this effectively can be not easy [81]. Companies that are successful with digital twins often emphasise the importance of change management, ensuring that teams understand the twin is a tool to aid, not replace, their decision-making. They also treat initial setbacks as learning experiences to refine the system.

Cost and ROI Concerns: The upfront investment for a comprehensive digital twin can be significant. Costs include sensors and IoT infrastructure, data storage and computing (mainly if big data or cloud services are utilised), software licensing or development costs for simulation/analytics platforms, and the labour costs of development and integration. For small and mid-size manufacturers, these costs can be a significant barrier. They need to be convinced of the return on investment (ROI). While, in theory, the twin will save money through efficiency and uptime gains, those gains may not be realised immediately, and the ROI may take time. Some companies are hesitant to proceed without clear benchmarks of success. Additionally, maintaining a twin incurs ongoing costs, including data bandwidth, cloud computing bills, and keeping models up to date, among others. It's worth noting, however, that as the technology matures, emerging solutions (including vendor-provided digital twin platforms and templates) are becoming available, which can lower the cost of entry. Still, each implementation tends to be at least partially custom, so management must be willing to invest with a strategic long-term view. One way companies address this is by focusing on a specific high-value use case (like a twin for the most critical machine that causes the most downtime) to prove value, then expanding from there.

In summary, while digital twins promise significant benefits, they also present a multifaceted challenge for manufacturing organisations. It requires careful planning, the right expertise, and often a willingness to collaborate on developing standards and best practices. Encouragingly, many of these challenges are being actively addressed by industry groups and SDOs (Standards Development Organisations). Manufacturers can leverage lessons learned by early adopters, engage in consortia (to share knowledge and even software tools), and adopt standards to mitigate some challenges[81]. Over time, as tools become more plug-and-play and standards improve interoperability, the barriers to entry are expected to lower. However, for now, any company embarking on a digital twin initiative should be mindful of these limitations and plan accordingly (for instance, ensuring strong data governance for data quality, involving IT security from the outset, and starting small while designing for scale).

Real-World Examples and Case Studies

Digital twin technology is being actively deployed by industry leaders in various manufacturing sectors. Below are several real-world examples and case studies illustrating how aerospace, automotive, and precision manufacturing companies are leveraging digital twins, along with the realised benefits:

Aerospace (Airbus): Airbus has embraced digital twins across the design, production, and operation of its aircraft. In design, they utilise high-fidelity twins of aircraft systems to run simulations (e.g., virtual stress tests and aerodynamic tests), which reduces the need for physical prototypes and accelerates development [77]. In manufacturing, Airbus creates digital twins of assembly lines and production processes. For example, when refurbishing an A380 facility for A321 production, they built a virtual model of the new assembly line to simulate product flow and optimise the layout before actual installation [77]. This ensured an efficient setup and helped identify bottlenecks in advance. On the shop floor, Airbus utilises "industrial digital twins" that aggregate machine data to monitor operations in real-time. At facilities like Hangar 9 in Hamburg, the twin-track production progresses against the plan, and in a helicopter gearbox production line, it automatically monitors the status of each station [77]. These twins have improved production agility and transparency – any delay or issue is immediately visible digitally, allowing managers to respond quickly and effectively. For quality control, Airbus's twins gather data from drilling and milling machines to detect quality deviations in aircraft parts. If a hole drilling operation starts to drift out of tolerance, the twin flags it, allowing corrections to be made and thereby maintaining high quality [77]. Furthermore, once planes are in service, Airbus utilises digital twins (via its Skywise data platform) to support predictive maintenance for airlines. Real-time flight and sensor data from over 12,000 in-service aircraft feed their digital counterparts, allowing Airbus and airline engineers to predict component wear and optimise maintenance schedules [77]. This has led to improved aircraft availability and reduced maintenance costs for operators. Overall, Airbus credits digital twins as "a cornerstone of our digital transformation... building each aircraft twice: first in the digital world, and then in the real one" [77], underscoring the central role the technology plays in their operations.

Aerospace (Rolls-Royce jet engines): (Another aerospace example is Rolls-Royce's use of digital twins for jet engine maintenance. Rolls-

Royce creates a digital twin for each engine in service, updated with sensor data from the engine's operation. This twin tracks the engine's performance and health parameters (vibration, temperature, pressure readings of various components) and is used to predict maintenance needs. By analysing deviations in the digital twin, Rolls-Royce can perform condition-based maintenance, scheduling an engine for service exactly when needed rather than at fixed intervals. This has improved reliability and reduced downtime for airlines using their engines[88]. [Rolls-Royce reference could be cited if available])

Automotive (Ford Motor Company): Ford has applied digital twin technology in vehicle design and factory planning. On the product side, Ford's use of virtual vehicle twins to run simulations (for crash tests, aerodynamics, etc.) has enabled them to reduce the need for physical prototypes. According to industry reports, Ford achieved a 25% reduction in vehicle design time by leveraging digital twins and simulation in the design phase[84]. On the manufacturing side, Ford utilises digital twins of its assembly lines to simulate the introduction of new models. For instance, before retooling a factory for a new car model, they create a twin of the modified line to test how the new process will perform, optimising workstation layouts and robot coordination virtually. This practice has minimised production ramp-up issues and ensured smoother launches.

Automotive (BMW): BMW is known for its "digital factory" initiatives. They have built digital twins of entire production plants to simulate and optimise everything from material flow to workforce logistics. In one reported success, BMW improved production efficiency by 30% through implementing virtual simulations of its manufacturing systems[84]. This involved using twins to experiment with various assembly sequences and logistics scenarios in a virtual environment, resulting in a more efficient workflow in the real world. BMW also utilises digital twins for robotics in manufacturing – each robotic cell may have a digital twin used for offline programming and error detection. By validating robot programs in the first place, BMW reduces the downtime needed for physical trial-and-error programming. The net effect is faster time to full production and higher throughput.

Automotive (Nissan): Nissan utilised digital twins to reduce prototyping costs significantly. By using virtual twins of new car designs and manufacturing processes, they achieved a 40% reduction in the number of physical prototypes, resulting in significant cost savings [84]. The twins allowed engineers to virtually test the fit and finish of components, assembly procedures, and even ergonomics for workers, ironing out issues before building physical test units. This not only saved money but also shortened development cycles for new models.

Automotive (Volkswagen & GM): Volkswagen implemented digital twins for real-time monitoring of its production lines. In one case, a twin helped consolidate data from various machines on the line to provide a unified live view and analytics (e.g., cycle times, downtime events). As a result, they saw a 20% decrease in unplanned downtime by reacting faster to issues and identifying root causes via the twins' analytics [84]. General Motors, on the other hand, focused on quality analytics – using digital twin models of its manufacturing processes combined with AI to predict quality issues. GM reportedly improved quality metrics by 15% using these predictive insights [84], meaning fewer defects and warranty claims. These examples show how even established processes in automotive manufacturing can be fine-tuned with twin technology.

Precision Manufacturing (CNC Machining example): A precision manufacturer (e.g., in aerospace component machining or medical device fabrication) might employ a digital twin of a machining process to ensure tight tolerances are met. One case study examined a digital twin for a multi-stage automotive production line that included CNC machining. The twin continuously monitored sensor data and adjusted processes, resulting in a 6% efficiency increase and dramatically reduced downtime, as noted earlier [82]. In a precision context, the twin can model tool wear and thermal expansion effects in machining with high accuracy. For example, a digital twin of a high-precision milling machine can track how slight temperature changes or tool wear affect part dimensions, and it can dynamically tweak cutting parameters or trigger compensation measures. This leads to better consistency in producing parts that are within very tight tolerances. Companies that manufacture jet engine components or precision optics have utilised digital twins to maintain quality without slowing down production. Another application in precision manufacturing is the use of digital twins for metrology and calibration. By comparing real-time production data with the twin, the system can automatically calibrate equipment (like adjusting a machine's alignment if the twin detects drift). The metrology news industry source highlights that digital twins help manufacturers maintain high quality by monitoring and controlling processes in real-time, essentially acting as an always-on quality control system [89].

These case studies demonstrate that digital twins are not merely theoretical concepts but practical tools that deliver real value. Table 1 provides a summary of selected examples across industries and their outcomes:

Industry & Company	Digital Twin Application	Outcomes / Benefits
Aerospace – Airbus	Virtual twin of aircraft design, production line, and in-service operation[77].	Reduced physical prototypes; optimised assembly line layout; real-time production monitoring; early quality deviation detection; predictive aircraft maintenance[77].
Automotive – Ford	Vehicle digital twin for virtual testing; Twin of an assembly line for new model introduction [84].	25% reduction in design time for new vehicles; smoother factory retooling with fewer launch issues[84].
Automotive – BMW	Plant-wide digital twin simulations for production optimisation [84].	A 30% increase in production efficiency was achieved by debottlenecking workflows and optimising robot coordination [84].

Automotive – Nissan	Product and process twins to reduce the need for prototypes [84].	40% reduction in prototyping costs; faster development cycles[84].
Automotive – Volkswagen	Real-time digital twin of the production line for monitoring [84].	A 20% decrease in unplanned downtime through quick issue detection and resolution [84].
Automotive – General Motors	Predictive quality analytics twin in manufacturing[84].	15% improvement in product quality control (defects reduced by anticipating issues)[84].
Precision Manufacturing – CNC Line	Twin of a multi-stage CNC production process (high precision machining)[81].	~6% increase in efficiency; significant reduction in downtime (~87%); consistently meeting tight tolerances via real-time adjustments[81, 82].

Table: Examples of digital twin applications in manufacturing and their benefits.

These examples illustrate that whether it's aerospace's complex products, automotive's high-volume production, or ultra-precision machining, digital twins are adaptable and beneficial. The common thread is leveraging data and simulation to gain insights that translate into improvements in efficiency, quality, and cost. Industry leaders often start with specific pilot projects (as seen with the targeted improvements at particular plants or processes) and then scale up once benefits are proven. It's also evident that partnerships sometimes drive these successes – many of the above cases involve collaboration with technology providers (for example, Airbus working with Dassault Systèmes for their 3DExperience platform[77], or OEMs partnering with simulation software companies). This ecosystem approach helps companies overcome specific challenges by not attempting to do everything in-house.

Enabling Technologies Supporting Digital Twins

Several advanced technologies converge to make digital twins possible and powerful in the manufacturing sector. The digital twin is not a standalone tool; it is enabled by the broader Industry 4.0 technology stack, which includes IoT for data collection, AI/ML for analytics, and cloud/edge computing for processing and storage. Key enabling technologies include:

Internet of Things (IoT) and Sensors: IoT serves as the foundation for data acquisition in digital twins. In manufacturing, IoT devices (sensors, actuators, and intelligent machines) continuously collect real-time data from the physical environment and feed it to the digital twin [90]. For example, vibration sensors on a motor, temperature sensors in a furnace, or RFID tags tracking materials in a factory all provide the raw streams that update the twin. Without IoT, a twin would be blind and static. Advancements in Industrial IoT (IIoT) have made it feasible to instrument nearly every aspect of a production line. Sensor costs have decreased, and wireless connectivity (such as Industrial Wi-Fi and 5G) has improved, enabling pervasive data collection. Protocols and standards such as OPC UA, MQTT, and Ethernet/IP facilitate the reliable transfer of sensor data from the shop floor to IT systems. IoT also encompasses actuators and control devices, which can receive commands from a digital twin analysis (closing the loop for adjustments). The proliferation of IoT means a twin can encompass diverse data, from machine states to environmental conditions (such as humidity and vibration on the factory floor), providing a holistic view necessary for accurate modelling. In essence, IoT provides the real-time telemetry and control hooks that a digital twin uses to stay synchronised with its physical twin and effect changes.

Artificial Intelligence and Machine Learning: AI/ML technologies are crucial for analysing the vast amount of data that digital twins collect and for developing predictive models. Machine learning algorithms can be trained on historical sensor data to detect patterns or anomalies that humans might miss. For instance, ML models can learn the normal operating vibration signature of a machine and flag deviations (predictive maintenance), or optimise process parameters by correlating myriad factors (such as temperature, speed, and material properties) with yield. AI techniques are also used to create surrogate models that run faster than physics-based simulations – for example, a neural network might approximate a complex physics process within the twin, enabling real-time simulation where a complete physics computation would be too slow. Predictive analytics, powered by ML, is at the heart of many digital twin benefits (like forecasting failures or quality issues before they happen)[81]. Additionally, AI enables the twin to provide prescriptive insights: using reinforcement learning, a twin might even suggest optimal control strategies in real time. Computer vision (an AI subfield) can be used in twins for quality inspection by analysing images of products and comparing them to ideal CAD models. As manufacturing twins generate large amounts of data, advanced analytics become indispensable for extracting actionable insights. AI is the set of techniques that makes this possible, transforming raw data into diagnostics, prognostics, and optimisations. In practice, companies often use AI platforms (like cloud ML services or specialised toolkits) integrated with the twin. For example, a twin might stream data to a cloud AI service that returns predictions (e.g., remaining useful life of a component), which the twin then uses to trigger maintenance workflows. Without AI/ML, a digital twin would largely be limited to raw monitoring and deterministic simulation; with AI/ML, it becomes predictive and adaptive, learning and improving its accuracy over time[90].

Cloud Computing: The cloud provides the scalable infrastructure that many digital twins require. Manufacturing twins can generate massive amounts of data (a large factory can produce terabytes of sensor data weekly). Storing and processing this data on-premises can be costly and inflexible. Cloud platforms (offered by providers like AWS, Microsoft Azure, Google Cloud, etc.) allow manufacturers to offload data storage and heavy computations to data centres that scale on demand. This is particularly useful for running complex simulations or AI models that require significant compute power. For example, simulating a whole factory or running computational fluid dynamics as part of a twin can be spun up on cloud HPC resources. Cloud also supports centralised data lakes where all the twin-related data (from multiple plants or lines) can be aggregated for enterprise-level analytics. Moreover, cloud services often come with IoT integration, analytics, and visualisation tools out of the box (for example, AWS IoT TwinMaker or Azure Digital Twins service), accelerating development. Collaboration is another benefit: cloud-based twins can be accessed by experts and stakeholders from anywhere (which proved helpful, for instance, during the COVID-19 pandemic when remote monitoring of factories became essential). However, latency can be an issue for time-sensitive tasks, which is why many architectures employ a hybrid approach that combines cloud and edge computing. Still, for batch analysis, machine learning training, long-term data archiving, and multi-site integration, cloud computing is a key enabler. It effectively provides a platform for deploying digital twin applications at scale, without requiring each company to build its own data centres.

Edge Computing: While cloud is great for scale, edge computing addresses the need for low-latency, on-site data processing for digital twins. Edge computing means placing computing resources closer to the source of data (on the factory floor, in the device, or a local server at the plant). For digital twins that require real-time control or rapid feedback (in the order of milliseconds to seconds), sending data to and from the cloud may be too slow or unreliable (if connectivity is lost). Thus, edge devices can run critical parts of the digital twin model locally. For example, an edge gateway at a factory might run an AI model that detects an anomaly in a machine's vibration within milliseconds and sends a shutdown command to prevent damage, all locally, without relying on the cloud [80]https://medium.com/@alex.brewer/revolutionizing-manufacturing-with-edge-computing-and-digital-twins-02ebbbd43850. Edge computing also helps filter and preprocess data before sending it to the cloud twin, reducing bandwidth usage. In some cases, a distributed digital twin architecture is employed: part of the twin (data collection and immediate response logic) is located on the edge, while another part (heavy analytics and long-term planning simulations) resides in the cloud. Modern IoT platforms support such hybrid deployments. The combination of edge and cloud ensures that a digital twin system can be both responsive and comprehensive – immediate actions handled at the edge, and big-picture analysis in the cloud. With the rise of 5G private networks in factories, edge devices become even more capable, as high-speed wireless links can connect many devices to a local edge server. Companies like Litmus and FogHorn (industrial edge platform providers) specifically emphasise edge computing as crucial for digital twin deployments in manufacturing, citing improvements in response times and reliability[80, 91]. Simply put, edge computing empowers devices to interact with the physical world in real-time, which is often necessary for control applications.

Simulation and Modelling Software: At the heart of any digital twin is the model. A variety of simulation tools and modelling approaches enable twins to mimic physical processes. This includes physics-based simulation software, such as CAD/CAM and CAE tools (Computer-Aided Design/Manufacturing/Engineering), used to create 3D models and run engineering analyses (stress, fluid flow, kinematics, etc.). These tools supply the base models for twins. For example, a finite element analysis (FEA) model of a machine component can be incorporated into a twin to simulate its physical behaviour under load. Similarly, computational fluid dynamics (CFD) models might be part of a twin for a casting or injection moulding process to predict material flow. Specialised factory simulation software (like Siemens Plant Sim, AnyLogic, or Dassault's DELMIA) can create discrete event or agent-based simulations of production lines, which can be tied to live data in a twin scenario. The challenge historically was that simulations were offline; what's changing with digital twins is that these simulations are increasingly kept in sync with live data. Co-simulation frameworks are sometimes used to integrate multiple models (mechanical, electrical, and software) to represent a complex system. Additionally, emerging modelling standards such as Modelica (for multi-domain physical modelling) or the FMU/FMI (Functional Mock-up Interface) standard enable different simulation models to interoperate, which can help build a composite twin from multiple sub-models. Vendors like Siemens, ANSYS, and PTC have developed digital twin solutions that integrate simulation with IoT data. For instance, ANSYS Twin Builder allows users to create a twin by linking physics models with IIoT inputs. These tools are vital in that they provide the virtual environment in which a twin can run experiments.

Advances in simulation fidelity and speed, including the use of GPUs for simulation (as Nvidia is doing with its Omniverse platform), directly enhance digital twin capabilities.

Augmented Reality (AR) and Virtual Reality (VR): While not strictly necessary for a digital twin to function, AR and VR are powerful interface technologies for interacting with digital twins. In manufacturing, AR can be used on the shop floor to overlay digital twin data onto physical equipment through tablets or smart glasses. For example, a maintenance technician wearing an AR headset might see the live pressure readings from a pump (coming from its digital twin) hovering above the physical pump, along with indicators of which part is predicted to fail next. This makes the twins' insights immediately accessible in context. Airbus mentioned using connected devices, such as tablets and smart glasses, to provide virtual training for operators, effectively allowing operators to practice on the digital twin of equipment in an immersive way [77]. VR, on the other hand, will enable engineers to step into a fully virtual factory or product. This is extremely useful during design and layout planning: one can virtually walk through a production line twin to inspect it for safety or efficiency issues before anything is built. It also facilitates collaborative design reviews – multiple stakeholders can meet in a virtual twin environment (even if they are remotely located) to explore improvements. The combination of twins with AR/VR is sometimes referred to as an aspect of the "industrial metaverse." It brings the twin to life in a human-accessible manner. Companies that leverage this approach have found that it improves understanding and speeds up training and troubleshooting, as humans can intuitively see how the system behaves. For instance, a complex machine's twin in VR can be used to train new operators on how to handle various scenarios (like jams or failures) safely. While AR/VR are not mandatory for a digital twin project, they significantly enhance user interaction with the twin, driving value by improving the decision-making process and human-system integration.

In summary, digital twins sit at the nexus of multiple modern technologies. IoT provides the nervous system (senses and action), AI provides the brain (intelligence and learning), cloud/edge computing provides the computing muscle, and advanced modelling software offers the body or structure of the twin. Together, this tech stack has reached a level of maturity that makes digital twins feasible at scale today – something that wasn't possible a decade ago when sensors were scarcer and AI less developed. Industry 4.0's progress thus directly feeds the rise of digital twins. For industry professionals, understanding these enabling technologies is essential not only for implementing twins but also for maintaining them. For example, ensuring your IoT devices are calibrated and your AI models are retrained as needed is part of sustaining an effective digital twin deployment[90]. Many companies adopt a platform approach, utilising IIoT platforms (such as PTC ThingWorx, Siemens MindSphere, or Azure IoT) that combine several of these technologies, offering sensor connectivity, data storage, and some analytics as a starting point. They then customise the twin logic for their specific needs. The good news is that the tools are becoming more accessible; the challenge remains in bringing them together in a purposeful way to solve real manufacturing problems.

Standards, Frameworks, and Emerging Trends

The growth of digital twins in manufacturing has spurred the development of standards and frameworks to ensure interoperability and guide best practices. At the same time, several emerging trends are shaping the adoption of digital twin technology and its future direction.

Evolving Standards and Frameworks for Digital Twins in Manufacturing

To tackle the challenges of interoperability and to accelerate adoption, international standards bodies and industry consortia are actively working on digital twin standards:

ISO 23247 – Digital Twin Framework for Manufacturing: ISO (International Organisation for Standardisation) released the ISO 23247 series (in 2021) specifically as a four-part standard for manufacturing digital twins[81]. This standard provides a generic framework that can be specialised for different manufacturing sectors (e.g., discrete, process)[92]. It defines key terms and a reference architecture, including functional entities needed in a manufacturing twin system. For example, it delineates layers such as sensing/observables, data communication, digital twin models, and user applications – ensuring a common taxonomyap238.orgap238.org. The goal is to make different implementations more compatible by adhering to the same architectural blueprint. ISO 23247 also emphasises the concept of the digital thread (connecting digital twins across the product lifecycle) and outlines how to manage data in a twin scenario. By following ISO 23247, a manufacturer can better structure their twin development and increase the likelihood that components from multiple vendors will interoperate. Early studies have shown that while adoption of ISO 23247 is in its initial stages, experts believe aligning with its reference architecture is "pivotal for tackling challenges such as interoperability and evolvability"[81]. Ongoing work in ISO is looking at extending this framework (e.g., proposed parts on digital twin data management and digital twin composition)[81].

Industrial Internet Consortium (IIC) and Digital Twin Consortium (DTC): The Industrial Internet Consortium (now incorporated into the Industry IoT Consortium) and the Digital Twin Consortium (launched in 2020 under the Object Management Group) are two influential industry consortia. The Digital Twin Consortium (DTC) brings together industry, government, and academia to define common digital twin terminology, reference architectures, and an open-source approach. They have published definitions (such as the one cited above for digital twin as a virtual representation synchronised with reality[86]) and white papers on interoperability frameworks[93]. DTC's approach is cross-industry, with a primary focus on the manufacturing sector. They work on use case reference architectures (e.g., a reference model for a factory twin) and integration with other standards, such as BIM (for buildings) and PLM systems. The consortium also fosters liaisons between standard bodies (for instance, a partnership with buildingSMART to align building and manufacturing twin standards). Meanwhile, the IIC published guidance like the "Digital Twin Interoperability Framework", which outlines how different digital twins can communicate in an IoT context[93]. It also introduced conceptual models such as the Digital Twin Capabilities Periodic Table (a framework categorising twin capabilities). For manufacturers, these consortium outputs serve as best practice guides. They can be used to verify that their digital twin implementations cover all necessary aspects (communication, data governance, etc.) in a standardised manner.

IEEE Standards: IEEE has been working on standards related to digital twins and smart manufacturing. One notable effort is IEEE P2806 (approved as a project in 2019), which aims to define a "System Architecture of Digital Representation for Physical Objects in Factory Environments." This standard outlines the structure of a digital twin for factory assets, including data modelling and interface standards. IEEE P2806 provides standardised terminology and architecture for digital twins, intended to support the development of "digital factories" by ensuring that all components of a twin (sensors, models, etc.) can fit into a typical architecture. Additionally, IEEE has other initiatives (like P2671/P2672 under IEEE Digital Transformation in manufacturing working groups) focusing on aspects such as digital twin maturity models and knowledge graphs for manufacturing. While ISO 23247 is an international consensus, IEEE's work often complements it by providing more technical detail or alternative approaches that can eventually be harmonised. For example, an IEEE P2806-based architecture can be seen as one implementation path that aligns with the ISO framework. As these standards finalise (P2806 might soon become an official standard if not already), manufacturers can reference them when designing systems to ensure they meet emerging global criteria for what constitutes a robust digital twin system.

OPC Foundation and Asset Administration Shell (AAS): The OPC UA standard is widely used for industrial interoperability (especially for machine-to-software communication). Recognising the importance of digital twins, the OPC Foundation, in collaboration with organisations such as Plattform Industrie 4.0, is working on standards to represent digital twin information models. One concrete development is the concept of the Asset Administration Shell (AAS) – a standardised digital representation of an asset defined in the Industry 4.0 context. The AAS can be thought of as a container for all digital information about an asset, which naturally aligns with the idea of a digital twin. Efforts are underway to align OPC UA's information modelling capabilities with the AAS, so that each machine or device's digital twin can be represented as an AAS accessible over OPC UA. ISO and IEC have been looking into making AAS part of the digital twin standards (IEC PAS 63088 references the AAS, for example). The benefit is that if all equipment provides an AAS, a digital twin platform could ingest those directly, greatly simplifying integration. OPC UA is also extending into the cloud domain (with the OPC UA Cloud Library) to allow easier sharing of models. For manufacturers, following these developments means their digital twin systems could more easily plug-and-play devices from different vendors as long as those vendors supply an AAS/OPC UA interface that adheres to the standard.

ASME Verification & Validation: As mentioned under challenges, ASME (a professional engineering organisation) has taken an interest in digital twin credibility. They formed a Digital Twin Task Group and have initiatives on VVUQ (Verification, Validation, and Uncertainty Quantification) for models used in digital twins[87]. This is more of a best-practice framework than a standard – it provides guidelines on how to test and validate a digital twin's accuracy. ASME's involvement is significant because many manufacturing companies rely on ASME standards for safety and quality; having ASME-endorsed practices for digital twins might become part of certification or compliance in the future (e.g., if a digital twin is used to certify a part's quality, one would need to show it was built and validated to ASME guidelines).

Others: There are many other sector-specific or regional efforts (for instance, ISO/IEC JTC1 is working on a digital twin reference architecture in the IoT context[81], which might be more IT-focused; ETSI in telecom is exploring digital twin for networks; GAIA-X in Europe touches on data spaces that could include manufacturing twin data, etc.. In Germany, the Industrial Digital Twin Association (IDTA) was established to develop the AAS concept further and promote its adoption in industry, effectively standardising the structure of digital twin data in alignment with Industrie 4.0 goals.

The existence of these standards and frameworks is a positive sign, as it indicates that the community is actively addressing the fragmentation issue. For an industry professional, staying informed about these standards is essential when evaluating vendors or solutions. Adhering to standards can future-proof a digital twin implementation. For example, choosing a platform that supports ISO 23247 or OPC UA AAS may ensure easier expansion and integration down the line. Additionally, compliance with standards can increase trust (internally and with customers) that the digital twin is built on solid foundations. Many companies are now including language in RFPs (requests for proposals) that asks for "ISO 23247 compliance" or "must align with Digital Twin Consortium definitions" to ensure that their investments aren't a proprietary dead end.

Emerging Trends and Future Outlook

Digital twin technology in manufacturing continues to evolve rapidly. Several emerging trends are influencing its adoption and shaping its future:

Digital Twins for Sustainability: Manufacturers are increasingly leveraging digital twins to achieve sustainability and energy efficiency targets. As noted, twins can optimise energy usage for machines and processes[81]. Beyond that, companies are now utilising digital twins at the facility or enterprise level to monitor their carbon footprint, simulate the impact of using alternative materials, or optimise logistics for lower emissions. A recent market analysis highlighted that a notable percentage of companies are deploying digital twins specifically to meet sustainability goals[85]. For example, a factory twin might be used to test scenarios for using solar power effectively on-site or reducing scrap waste. This trend aligns with broader corporate ESG (Environmental, Social, Governance) commitments. We can expect future digital twin solutions to have built-in modules for sustainability analytics – essentially "green twins" that focus on resource and energy optimisation. This not only benefits the environment but also reduces costs, making a strong business case.

Integration of AI and Advanced Analytics (Cognitive Twins): While AI is already an enabler, the trend is towards even smarter or cognitive digital twins. These are twins that not only predict but also prescribe and possibly self-optimise. With advances in AI, especially in areas like reinforcement learning and generative AI, future digital twins might autonomously experiment within themselves to find optimal process settings (acting like a digital process engineer). We also see AI being used for higher-level decision support: for instance, combining a digital twin with AI that can reason about scenarios in natural language, allowing users to query the twin in intuitive ways ("What happens if I increase machine X's speed by 10%?" and the AI-twin system can simulate and answer that). Some research projects are exploring the merging of knowledge graphs (to encode expert knowledge) with digital twins, so that the twin not only has raw data but contextual knowledge to make decisions. As these capabilities grow, the role of human operators may shift more towards supervision, while the twin handles routine optimisation – a step towards more autonomous factories.

Edge-Heavy Deployments and 5G: The rollout of 5G networks and improved industrial connectivity is enabling more robust edge computing for digital twins. 5G's low latency and high device density mean even more sensors can be deployed and reliably connected, and computations can be distributed. A trend is emerging where critical control twins (e.g., for a robotic cell) run entirely on the shop floor edge, sometimes directly on the PLC or industrial PC controlling the cell, to achieve ultra-fast response. At the same time, the cloud is used for coordination between cells and long-term analysis. Factories are adopting private 5G to connect thousands of devices (robots, AGVs, tools) with guaranteed QoS; these networks will serve as the data highways for rich digital twin data. This paves the way for real-time digital twin systems at scales previously not possible (imagine an entire automotive plant with every robot and conveyor streaming data in real-time into a plant twin – 5G makes that far more feasible in terms of bandwidth). Thus, the twin paradigm will benefit from the continued evolution of industrial connectivity and edge hardware.

Interoperability and Digital Twin Marketplaces: As standards become established, we anticipate the emergence of digital twin marketplaces or ecosystems. This is where third-party vendors can offer pre-built twin models of common equipment that can be plugged into a company's digital twin platform. For example, a machine manufacturer might supply a high-fidelity digital twin model of their machine as a selling point, and the customer can integrate that into their factory twin easily (if both follow standard interfaces). This concept is bolstered by interoperability initiatives (Trend: collaborative ecosystems). IoT Analytics reported that partnerships between industrial OEMs and cloud providers/simulation specialists are on the rise[85] – e.g., an automation company teaming with a cloud platform to provide twin templates for customers. Shortly, if a factory buys ten new machines, it might also receive digital models of those machines, which it can instantiate in its twin system with minimal effort. Additionally, marketplaces could allow sharing/selling of analytics apps for twins (e.g., an app for predictive maintenance that can be applied to your twin). This type of ecosystem will accelerate twin adoption, as not everything needs to be custom-built in-house.

- Human-Twin Interaction and Workforce Enablement: There is a trend towards making digital twins more accessible to the workforce on the factory floor. We touched on AR/VR for training; going forward, more intuitive interfaces (voice queries to the twin, AR overlays in daily work) will emerge. Digital twins might also incorporate more human behaviour modelling – for example, including a simulation of operator tasks in the overall process twin to optimise not just machines but also how people work with them (this can improve ergonomics and safety). As the workforce becomes more tech-savvy, having an operational twin that teams use daily (maybe each morning starts with a "twin dashboard" review of production health) might become standard. The twin becomes a communication tool bridging management and operators – everyone can see a common picture of the production status and issues, which can improve alignment and response.

- Expansion to New Domains (Beyond the Factory): We're also seeing digital twin concepts expanding beyond individual factories to multi-factory networks and supply chains, and even to product end-users. For example, some companies discuss a "digital twin of the organisation", which encompasses all assets, processes, and even employees represented digitally. While that's an ambitious notion, elements of it are happening – supply chain twins, logistics twins, and customer usage twins (for feedback into product design) are all being developed. In manufacturing, this means the digital twin won't stop at the factory walls: it will link upstream to suppliers (some companies are creating twins of their procurement and supplier quality processes) and downstream to distribution and customers, achieving a complete digital thread. This trend is facilitated by cloud connectivity and data sharing platforms. It also intersects with other concepts, such as predictive supply chain analytics and customer experience modelling. Essentially, manufacturing digital twins is a part of a larger digital ecosystem of an enterprise. The companies that capitalise on this may achieve the dream of proper end-to-end optimisation – design, manufacturing, and service all integrated via interconnected twins.

- Increased Adoption and ROI Proof Points: Lastly, the adoption trend itself is accelerating. As of 2023, roughly 29% of global manufacturing companies had at least partially implemented digital twin strategies (up from 20% in 2020)[85], and this number is growing. The digital twin market is projected to continue expanding at ~30% CAGR through 2027[85], indicating robust investment. With more adoption, we'll have more data on ROI, which creates a positive feedback cycle – success stories reduce hesitation for newcomers. We are likely to see digital twin technology trickling down from large enterprises to smaller manufacturers as solutions become more turnkey and affordable. Also, academic institutions are increasingly incorporating digital twin concepts into engineering and operations curricula, meaning the next generation of engineers will enter the workforce with familiarity in these tools, helping drive further implementation.

In summary, the future of digital twins in manufacturing is bright and dynamic. Standards efforts (ISO, IEEE, etc.) are addressing current pain points, aiming to make digital twins more plug-and-play and trustworthy. Emerging trends like sustainability focus, AI-driven autonomy, and seamless integration with both edge devices and enterprise systems are expanding the capabilities and value of digital twins. Industry professionals evaluating this technology should keep an eye on these trends: adopting a digital twin today means entering a rapidly evolving field. It's wise to design systems with flexibility, choosing open standards and scalable technologies, to be able to incorporate new advances (like next-gen AI or new standard protocols) when they arrive. The companies that leverage digital twins effectively stand to gain a competitive edge in efficiency, agility, and innovation, which in the high-pressure manufacturing sector can be the difference between leading the market or falling behind. As Airbus's experience and others show, digital twins are becoming an indispensable tool for manufacturing excellence. As tools and standards mature, their deployment will become increasingly routine across the industry.

MES in Manufacturing: Enabling Digital Transformation

Figure: The ISA-95 automation pyramid positions MES at Level 3 (Manufacturing Operations Management), bridging shop-floor control systems (Levels 0–2: sensors, PLCs, SCADA/HMI) and enterprise planning systems (Level 4: ERP)[94]. This hierarchy illustrates how MES connects real-time production operations with higher-level business processes.

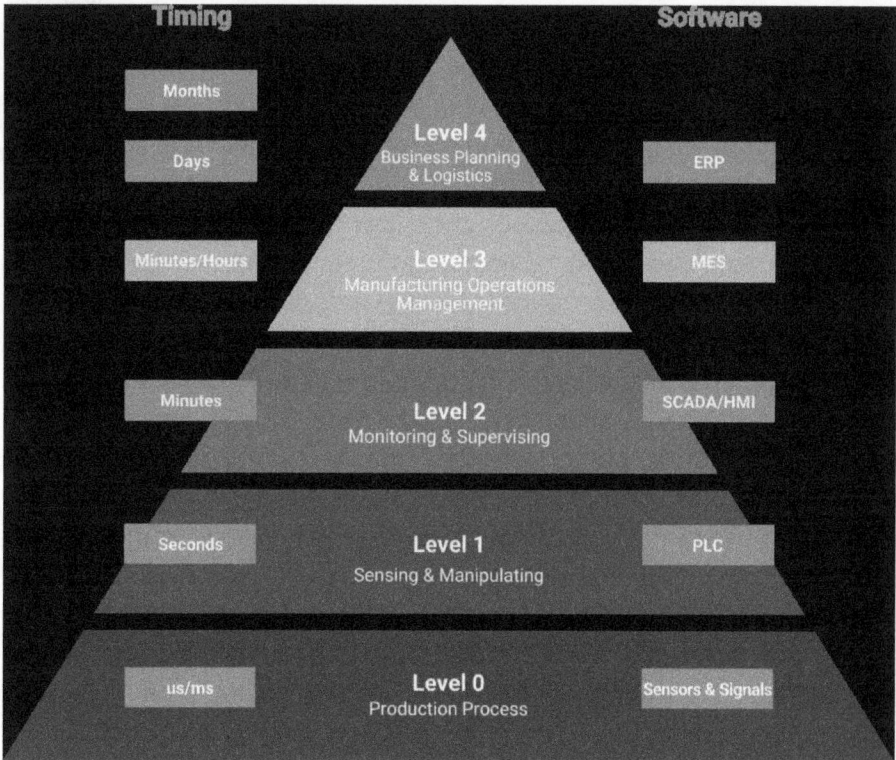

Timing		Software
Months		
Days	**Level 4** Business Planning & Logistics	ERP
Minutes/Hours	**Level 3** Manufacturing Operations Management	MES
Minutes	**Level 2** Monitoring & Supervising	SCADA/HMI
Seconds	**Level 1** Sensing & Manipulating	PLC
us/ms	**Level 0** Production Process	Sensors & Signals

A Manufacturing Execution System (MES) is a software-based solution that monitors, controls, and optimises production processes on the factory floor in real time[95]. In essence, an MES tracks the transformation of raw materials into finished goods, collecting accurate data at each step from order release to product completion[95]. By doing so, it provides manufacturers and decision-makers with a live view of work-in-progress (WIP), product genealogy, equipment status, and other operational details, enabling better control and timely adjustments to the production process[95]. The primary goal of MES is to ensure effective execution of manufacturing operations and improve overall production output (e.g. higher throughput, better quality, less waste)[95].

MES in the Technology Stack:

In the automation technology stack, MES occupies a central position between business planning systems and shop-floor control systems. Industry standards, such as ISA-95, define MES as the Level 3 layer of a manufacturing enterprise, situated between Level 4 enterprise systems (e.g., Enterprise Resource Planning, ERP) and the real-time control layers at Levels 0–2 (sensors, actuators, PLCs, SCADA) [95, 96]. In other words, MES is the functional layer that bridges the gap between ERP and the plant floor control systems, translating high-level production plans into detailed shop-floor instructions and feeding production data back up to business systems[95]. While an ERP system plans what and when to produce (handling orders, inventory, scheduling, logistics), the MES determines how to execute production on the shop floor with optimal efficiency [96]. At the same time, MES interfaces with process control and supervisory systems (like SCADA and distributed control systems) to gather real-time machine data and orchestrate workflows[97]. This intermediary role means that MES is often described as the "brains" of a smart factory, providing the execution muscle and real-time intelligence that links top-floor business decisions with shop-floor actions [97, 98]. In summary, MES fits within the manufacturing technology stack as the core layer for manufacturing operations management, ensuring that enterprise-level plans are executed efficiently on the factory floor, with complete visibility and control over the production lifecycle[95].

Core Functions of MES

Modern MES platforms are feature-rich, encompassing a broad range of functions necessary to manage and optimise production. The Manufacturing Enterprise Solutions Association (MESA) defined 11 core MES functions (known as the MESA-11 model), which, though refined over time, remain the foundation of most MES solutions[96]. These core functions include the following[96]:

Resource Allocation and Status: Managing the status and availability of manufacturing resources – machines, tools, materials, and labour – in real time. An MES tracks machine uptime/downtime, labour utilisation, and material consumption, allowing adjustments to resource assignments to meet production needs[96]. This ensures the right resources are in place to fulfil the production schedule.

Operations/Detailed Scheduling: Sequencing and timing of production tasks. MES takes high-level production schedules (often from ERP) and refines them into detailed, minute-by-minute schedules for each machine or work centre[96]. By optimising task sequencing based on priorities and real-time constraints (e.g., machine availability or changeovers), MES helps reduce waiting times and keeps work centres busy with minimal idle time.

Dispatching Production Units: Real-time dispatch of work orders and instructions to the shop floor. MES directs what job each workstation or operator should work on at any given time, based on the current schedule and any trigger events[96]. This function ensures a smooth workflow by communicating production orders and any changes immediately to the floor.

Document and Data Control: Handling of all production-related documents and data records. MES manages electronic work instructions, standard operating procedures, drawings, batch recipes, and any documents needed for production[96]. It ensures the latest versions are accessible to operators and that any process data (e.g. batch records, parameter values) is recorded and stored securely. By going paperless, MES reduces errors and makes data instantly available to all stakeholders[96].

Data Collection and Acquisition: Automatic collection of production data from machines, sensors, and operators as tasks are performed[96]. MES interfaces with shop-floor equipment (via sensors, PLCs, IIoT devices) to record process parameters, cycle times, quantities produced, temperatures, pressures, etc. This real-time data acquisition provides the raw information necessary for monitoring and analysing the production process.

Labour Management: Tracking and management of human resources in production. MES can log operator attendance, qualifications/certifications, work hours on each job, and even enforce that only authorised or trained personnel perform specific tasks[96]. This function helps ensure adequate staffing, tracks labour productivity, and can be tied to performance or training records.

Quality Management: Monitoring and enforcement of product quality procedures during manufacturing[96]. MES commonly integrates Statistical Process Control (SPC) checks, records quality inspection results, flags deviations or non-conformances, and can even halt production or trigger alerts if quality issues arise[96]. By capturing high-quality data in real-time, MES helps identify defects early, thereby reducing scrap and rework.

Process Management: Oversight of the end-to-end production process to ensure it flows according to plan and adheres to defined recipes or routings[96]. MES provides visibility into each step of manufacturing, identifies bottlenecks, and ensures adherence to process steps and routing sequences. This yields full traceability of the production process, highlighting points that affect throughput or quality.

Maintenance Management: While dedicated maintenance systems (CMMS) often handle detailed maintenance scheduling, MES contributes by using production data to predict equipment issues and coordinate maintenance with production[96]. For example, MES may track machine cycles or hours and alert when preventive maintenance is due, or it can create work orders for maintenance when certain conditions (such as temperatures or vibrations) exceed specified ranges. This minimises unplanned downtime by addressing issues proactively.

Product Tracking and Genealogy: Tracking the progress of each unit, batch, or lot through the factory and maintaining a genealogy record of the materials and processes it has undergone[96]. MES assigns identifiers to lots or serial numbers for units and records which raw materials (by lot/serial number) and components went into each product, as well as the process conditions and test results at each step. This comprehensive traceability is crucial for regulated industries – it enables rapid root-cause analysis and targeted recalls by linking finished goods back to specific material lots and process history[96].

Performance Analysis: Analysing production results and comparing them against goals or key performance indicators (KPIs)[96]. MES typically computes metrics like throughput, yield, downtime, and Overall Equipment Effectiveness (OEE) to gauge how well the plant is running[98]. It can generate reports and dashboards that highlight inefficiencies or trends, providing a basis for continuous improvement in the manufacturing process.

These functions collectively enable an MES to orchestrate production and provide a single, integrated source of truth for manufacturing operations. By digitising production activities and data, MES eliminates manual paperwork, improves data accuracy, and ensures that every step, from scheduling to execution to quality control, is managed within one coherent system. Not all MES implementations will utilise every module extensively; companies often start with a subset of MES functions that are most critical to their needs (for example, focusing first on production scheduling and genealogy in a highly regulated environment). However, the real power of MES lies in its ability to unify these functions and serve as the nerve centre for all production activities.

MES Integration with ERP, PLM, SCADA, and Shop-Floor Equipment

One of the defining characteristics of an MES is how it interfaces with other enterprise and industrial systems. MES does not operate in isolation; it is the connecting hub that ensures data flows seamlessly between business-level planning systems, engineering systems, and the automation on the factory floor. Key integration points include:

ERP ↔ MES: The integration between ERP (Enterprise Resource Planning) and MES is critical for a closed-loop manufacturing environment. The ERP (Level 4 in ISA-95) manages business processes such as order management, materials purchasing, inventory tracking, and production planning. MES (Level 3) receives production orders, work orders, and schedules from the ERP and executes them on the shop floor[96]. As production progresses, MES sends updates back to the ERP – for example, completed job statuses, actual production quantities, material consumed, and any quality dispositions. Together, an ERP-MES duo provides a holistic view: ERP determines what products to make and manages resources (materials, finances, etc.), while MES determines how to make those products efficiently and with quality, given real-time conditions[96]. When properly integrated, ERP and MES function as a unified ecosystem, where MES executes the ERP's plans with feedback loops that update inventory and schedule information in real time[96]. This reduces manual data entry and ensures that enterprise-level data (like inventory on hand or order status) is always up-to-date with what's happening on the factory floor.

PLM ↔ MES: Product Lifecycle Management (PLM) systems manage product design data, bills of materials (BOMs), recipes, and process definitions through the design and engineering phases of a product's life. Integrating PLM with MES creates a closed loop between design and manufacturing. Through this integration, engineering changes made in the PLM system (for instance, an update to a product's design or a change in a manufacturing process specification) can be automatically communicated to the MES, which in turn propagates those changes to the shop floor (e.g. updating work instructions or recipe parameters)[99]. Conversely, MES can feed production data back to PLM to inform engineers of how the product and process performed (e.g. defect rates, rework causes, cycle times), helping to drive design-for-manufacturability improvements. According to industry experts, integrating MES with both ERP and PLM enables a "single source of truth" for product and process data, where everyone, from design engineers to production managers, works off the same real-time information [99]. This closed-loop integration improves change management (ensuring that the latest specifications are always used in production). It fosters a digital thread that ties together product design, manufacturing execution, and performance data across the product's lifecycle.

MES ↔ SCADA/HMI and Control Systems: The MES connects with shop-floor control systems, including SCADA (Supervisory Control and Data Acquisition), HMI (Human-Machine Interface), DCS (Distributed Control Systems), and PLCs (Programmable Logic Controllers). While SCADA and control systems operate at ISA-95 Levels 1–2, directly interfacing with machinery to control loops, start/stop equipment, and gather sensor readings, MES operates at a higher level of coordination and analysis. SCADA is focused on real-time machine control and monitoring – it communicates with PLCs to execute control logic and logs data in historians, managing alarms and equipment states[97]. MES utilises data provided by SCADA (and directly from PLCs or IoT sensors) to perform production management functions, including tracking progress, analysing performance, and ensuring workflow adherence [97]. For example, a SCADA system might generate an alarm that a machine has tripped or produce a reading that a temperature is out of range; the MES can receive that event and respond by halting the production lot, triggering a quality hold, or dispatching a maintenance task. Essentially, SCADA is on the operational control side, and MES is on the production management side. By integrating the two, data flows from machine sensors up to the MES for analysis, and the MES can send commands or setpoints back down (indirectly via SCADA or directly to PLCs) to adjust processes [97]. This integration yields a more adaptive production environment: for instance, if MES detects via SCADA data that a process is drifting, it can enforce adjustments or alert operators before a quality issue occurs. Many modern platforms blur the line between MES and

SCADA; however, conceptually, MES converts the raw data from SCADA into actionable information and decisions for optimising production [97].

MES ↔ Shop-Floor Equipment and IoT: Beyond traditional SCADA, MES increasingly connects with innovative equipment and Industrial Internet of Things (IIoT) devices on the shop floor. This includes direct integration with CNC machines, robots, conveyors, sensors, and innovative tools that can report status or be controlled via standard protocols (like OPC UA, MQTT, etc.). By interfacing with equipment controllers, MES can collect granular data (such as machine cycle times, energy consumption, and vibration levels) and associate it with specific production orders or product units. It can also send instructions, recipes, or parameter sets to machines for automated setup and configuration. This direct integration is vital for both high-speed discrete manufacturing and complex batch processing – it closes the automation loop by ensuring the MES knows precisely what is happening on each machine at every moment. For example, in a pharmaceutical plant, an MES may communicate with a batch reactor's control system to download the recipe parameters for a batch and then capture all process data during execution for the electronic batch record. Or in electronics assembly, MES might interface with a placement machine to feed it the correct component loading list and then record which serial-numbered components were placed on each board. By integrating with IoT sensors, MES can also monitor conditions such as temperature, humidity, or tool wear in real-time, thereby improving quality control and enabling predictive maintenance.

In summary, MES serves as a central integration hub in a manufacturing IT/OT architecture. It provides a two-way bridge linking high-level enterprise systems (ERP, PLM, supply chain systems) with plant-floor technologies (SCADA, PLCs, IoT devices)[100]. This integration ensures that information flows seamlessly across what is often called the IT/OT divide – Information Technology systems in the office and Operational Technology on the factory floor[100]. When MES integration is successfully implemented, manufacturers achieve end-to-end visibility: an order entered in the ERP triggers actions and data collection in the

MES at the machine level, and any shop-floor event or result is promptly fed back to business systems and engineering systems. Such connectivity is essential for agile, responsive manufacturing, where changes in demand or design propagate quickly to production, and issues in production are immediately visible to stakeholders throughout the organisation.

Benefits of MES Implementation

Implementing a Manufacturing Execution System can yield significant benefits for manufacturing organisations. By digitising production management and providing real-time control, MES addresses many of the pain points of traditional, paper-driven manufacturing. Below are key benefits of MES implementation, as observed across industries:

Improved Productivity & Efficiency: MES increases production efficiency by coordinating workflows and reducing downtime. Real-time production monitoring enables the quick identification and resolution of bottlenecks or issues that could slow production [78]. For example, if a machine goes down or a workstation is starved of parts, the MES can reschedule tasks or alert personnel immediately, minimising idle time. MES also automates data collection and transactions that were previously performed manually (such as logging production counts or moving inventory), which speeds up the flow of information and keeps machines running efficiently. By optimising detailed schedules and dispatching, MES ensures that equipment and labour are utilised to the fullest extent, often improving OEE (Overall Equipment Effectiveness) on critical assets [78]. Some studies show double-digit percentage improvements in throughput or productivity after MES adoption, thanks to more synchronised and transparent operations.

Real-Time Visibility and Decision-Making: With MES, production managers and other stakeholders gain real-time visibility into the status of the factory floor. Every critical metric – current output, machine statuses, WIP inventory, order progress, and quality statistics – is accessible in real-time. This real-time visibility empowers better decision-making because managers base their decisions on up-to-the-minute data rather than after-the-fact reports[78]. For instance, MES dashboards may indicate that a specific production line is falling behind schedule; supervisors can then proactively add shifts or reallocate resources to bring it back on schedule. The MES acts as a central information hub, often providing alerts or notifications when key performance indicators deviate from the norm, enabling prompt action. In short, MES provides real-time manufacturing intelligence, allowing the operations to be managed based on facts rather than assumptions or delayed feedback.

Enhanced Quality Control: Quality is a significant focus of MES implementations. An MES enforces quality procedures by integrating in-process quality checks, data collection for critical quality attributes, and even stopping processes when out-of-spec conditions are detected[96]. Because quality data (measurements, test results, inspection outcomes) are recorded immediately and linked to each production unit, issues can be caught and corrected early. For example, if a dimensional check fails at an inspection station, MES can halt that lot from progressing, preventing a batch of defects from moving forward. By analysing quality trends, MES helps identify root causes of defects (such as a particular machine drift or an operator error) so that corrective action can be taken. Real-time quality monitoring reduces scrap and rework by ensuring only conforming products move to the next stage [96]. Over time, MES-driven quality data supports continuous improvement and more robust process control (e.g. enabling Six Sigma or statistical process control initiatives on the shop floor). Moreover, having all quality records in a centralised MES database simplifies compliance with standards and makes audits easier, since complete electronic records are readily available.

Regulatory Compliance and Traceability: MES greatly improves traceability and compliance, which is especially beneficial in industries with strict regulatory oversight (such as pharmaceuticals, medical devices, aerospace, and food and beverage). By recording the genealogy of each product – which materials were used, which processes and equipment it went through, who worked on it, and when – MES creates an electronic audit trail that meets regulatory requirements for production record-keeping[96]. For example, pharmaceutical manufacturers use MES to comply with FDA regulations (like 21 CFR Part 11 for electronic records and signatures) by capturing complete batch production records and enabling electronic signatures and validations. In the food industry, MES facilitates compliance with food safety regulations by ensuring every ingredient and process step is traceable, which is critical if a recall is needed[101]. The benefit of this comprehensive traceability extends beyond legal compliance to faster investigations of quality issues. If a defect or contamination is discovered, a manufacturer can quickly pinpoint affected lots and their distribution, thereby minimising risk to consumers and the business. Additionally, MES can enforce compliance by ensuring operators follow the proper processes (via electronic work instructions and sign-offs) and by integrating required checks (for example, verifying that a critical safety test is performed and recorded before a lot proceeds). This reduces the likelihood of non-compliance and the risk of costly recalls, rejections, or legal penalties.

Inventory Reduction and Better Inventory Accuracy: MES provides tighter control over WIP (work-in-progress) and inventory in production. Because the MES updates material consumption and production counts in real-time, inventory records are more accurate and up-to-date [96]. This accuracy enables companies to operate with lower buffer stocks; they don't need to keep as much "just-in-case" inventory because they have confidence in the visibility of what is on hand and its location. MES often integrates with or informs inventory management systems to automatically deduct raw materials as they are consumed on the line and to add finished goods as they are produced[96]. By reducing uncertainty, manufacturers can reduce excess WIP and safety stock, which in turn reduces carrying costs and waste. In some cases, MES also helps implement just-in-time (JIT) production by linking with demand signals from ERP and orchestrating material movements to the line exactly when needed. Overall, companies experience improved inventory turns and reduced cash tied up in inventory after deploying an MES that tightly coordinates production and inventory.

Shorter Lead Times and On-Time Delivery: By streamlining production processes and enhancing coordination, MES can reduce manufacturing cycle times, leading to improved on-time delivery performance. Detailed scheduling and synchronisation of operations ensure that orders flow through production with fewer delays. MES can also quickly adjust schedules in the face of disruptions (machine downtime, material shortage) to reduce the impact on delivery. The result is more predictable and shorter manufacturing lead times. Customers benefit because the company can deliver products more reliably on schedule. Internally, this also means increased agility to respond to changes in demand – MES makes it easier to shift production or expedite orders when priorities change, as the system can rapidly recalculate schedules and communicate changes to the floor. Studies have shown that plants with MES report higher adherence to schedule and faster response to order changes compared to prior manual scheduling methods[98].

Paperless Operations and Knowledge Retention: An often underrated benefit of MES is the move to a paperless (or largely paperless) shop floor environment. Traditionally, manufacturing involves a significant amount of paperwork, including work orders, batch records, checklists, and log sheets. MES digitises all of that, which means operators interact with screens or handheld devices to get their work instructions and enter data, rather than shuffling papers. This reduces errors caused by illegible handwriting, lost documents, or outdated paper instructions [96]. It also saves time – data entered once is available everywhere, and reports can be generated automatically rather than collating paper forms. Furthermore, going paperless with MES preserves organisational knowledge: process settings, best practices, and historical production data are all stored and can be searched or analysed, rather than being locked away in file cabinets. New employees can be trained more easily with on-screen instructions, and consistency is improved because everyone references the same digital instruction set. In regulated industries, electronic records reduce the time and effort of compliance (no more chasing signatures or archiving boxes of paper). Overall, the MES drives a more connected and transparent workplace, where information is available when and where it's needed.

Better Decision-Making with Analytics: With the rich dataset that MES captures (production rates, quality results, downtime reasons, etc.), companies can perform deeper analysis and continuous improvement. MES often includes reporting and analytics tools to calculate KPIs like cycle time, yield, OEE, mean time between failures, and so on[98]. These metrics allow management to identify performance gaps – for instance, highlighting that a specific production line has lower OEE due to frequent minor stoppages, or that one product has a higher defect rate than others. Armed with this information, improvement teams can focus their efforts where they matter and track the impact of changes over time. In essence, MES turns factory floor data into actionable insights, enabling data-driven decision making at all levels of operations[98]. Additionally, real-time data from MES can be shared enterprise-wide, so that other departments (like supply chain, engineering, or customer service) have visibility into manufacturing status and can make coordinated decisions (for example, adjusting procurement if a yield issue is slowing production of a key component).

In summary, the benefits of MES span productivity, visibility, quality, compliance, and agility. Many manufacturers report that after MES implementation, they have greater control over their operations and can respond more quickly to both problems and opportunities. It's important to note, however, that the degree of benefit realised depends on how well the MES is implemented and adopted on the shop floor. When fully embraced, an MES becomes the backbone of daily operations, driving a culture of discipline (following standardised procedures), accountability (measuring results in real-time), and continuous improvement (leveraging data to improve). These advantages collectively help manufacturers lower costs, improve customer satisfaction (through better quality and delivery), and gain a competitive edge in the marketplace[102].

Challenges and Limitations in MES Adoption

While the value proposition of MES is compelling, deploying an MES is a complex undertaking, and organisations often encounter significant challenges during the adoption process. Understanding these challenges is essential for industry professionals to plan effectively. Key challenges and limitations include:

Integration Complexity (Legacy Systems Compatibility): MES must integrate with a host of existing systems (ERP, legacy databases, home-grown applications, SCADA/PLC systems, etc.). This integration can be technically complex and time-consuming [103]. Many factories have older legacy systems or equipment that were never designed to connect to modern MES software. For example, a plant might have a 20-year-old production database, a custom scheduler, or machines that output data in proprietary formats. Interfacing these with a new MES can require custom middleware, drivers, or even upgrades of the legacy components. Incompatible data formats and communication protocols pose hurdles, often necessitating significant IT effort to bridge the gaps [103]. This complexity can prolong MES projects and increase costs. Additionally, without careful planning, integration can result in fragile connections that are difficult to maintain and sustain. Companies overcoming this challenge typically invest in robust integration middleware or platforms (such as OPC UA for equipment or API layers for business systems). They may enlist experienced system integrators to ensure that all parts of the enterprise communicate correctly with the MES. Following standards like ISA-95 (which defines data models for MES-ERP exchanges) can also mitigate integration issues by using a standard information model [96, 104].

High Implementation and Ownership Costs: Implementing MES software is a significant investment. There are the costs of MES licenses or subscriptions, the infrastructure costs (servers, networking, possibly new shop-floor terminals or tablets), and the service costs for configuration and customisation. Implementation projects can cost large enterprises hundreds of thousands or millions of dollars. Moreover, it's common for MES projects to exceed budget due to unforeseen challenges like more customisation than initially planned, or integration difficulties that require extra effort [103]. Beyond initial deployment, ongoing ownership costs include system maintenance, support contracts, and the need for internal or external MES expertise to accommodate process changes or upgrades. The cost factor can be a significant barrier, especially for small to medium-sized manufacturers. Justifying the ROI (return on investment) of MES often requires quantifying improvements in efficiency, scrap reduction, inventory savings, etc., which can be abstract before the system is in place. To control costs, best practices include phased implementations (starting with critical functions), avoiding excessive customisation (adapting business processes to the MES where possible, rather than the other way around), and ensuring strong executive sponsorship to keep the project on track. Even so, MES adoption is a strategic, long-term investment, and companies must be prepared to incur upfront expenditures in exchange for future gains.

Change Management and Workforce Adoption: Introducing an MES often entails significant cultural and process changes for the organisation. Production operators, technicians, and even management may be accustomed to specific ways of working (e.g. paper logs, manual decision-making, informal communication) that an MES will transform. This can lead to resistance from employees who are uncomfortable with new technology or fear that increased monitoring will be used punitively. User adoption is a common challenge – if operators do not use the system as intended (for example, neglecting to input data or bypassing MES procedures), the value of the MES diminishes greatly[103]. Additionally, some workers may feel that the MES imposes too rigid a structure or that it reduces their autonomy. Effective change management is crucial for addressing these concerns [103]. This includes communicating the benefits of the MES to all staff levels (how it will make their jobs easier, not harder), providing comprehensive training, and involving end-users early in the implementation process so that their feedback can be incorporated. It's also essential to create a support system – e.g. having MES "super-users" or champions on the shop floor who can help their peers and relay issues to the project team. Management must demonstrate a commitment to utilising the MES (for instance, by tracking KPIs from the MES rather than relying on legacy reports, which signals that the new system is the source of truth). Over time, as users become familiar with and see improvements (such as reduced paperwork or faster problem resolution), resistance tends to diminish. Still, companies should not underestimate the cultural shift MES may require, moving toward more data-driven and disciplined operations, which can be a difficult transition in the early stages.

Customisation and Implementation Complexity: Every manufacturing operation has unique aspects – special process steps, data to capture, or ways of organising production – which may not fit perfectly into an out-of-the-box MES solution. As a result, projects often require some degree of customisation or configuration to model the specific workflows and rules of the plant. Extensive customisation, however, can become a double-edged sword [103]. On one hand, it allows the MES to fit the business exactly; on the other, it increases project complexity, time, and cost, and can make future upgrades difficult (since custom code might need rework for new MES versions). There is a risk of "scope creep," where more and more custom features are added to satisfy every user demand, which can delay implementation and inflate budgets [103]. Striking the right balance is a challenge: organisations must distinguish between processes that truly provide a competitive advantage (and thus justify customisation) and those that could be adjusted to standard MES best practices. Implementing an MES often forces companies to evaluate and possibly re-engineer some of their processes to align with the software. This can be challenging in itself, as it may disrupt established routines. Adopting the MES's standard functionalities wherever possible can speed up deployment, but it requires flexibility on the part of the business to adapt to new ways of working. Many MES vendors and implementers have industry templates or best-practice models; leveraging these can reduce the need for customisation. Nonetheless, the complexity of implementation remains a hurdle, requiring strong project management, precise requirements, and close collaboration between the manufacturing engineers, IT, and the MES provider.

Legacy Infrastructure and Data Migration: When introducing MES into a plant that has been running for decades, a practical issue is how to handle existing data and systems. Legacy production records, recipes, or work instructions might exist in various formats (paper files, spreadsheets, outdated databases). Deciding how much historical data to migrate into the MES and how to do it can be a challenge. Migrating data carries the risk of errors and can consume significant effort (cleaning and validating old data). Additionally, legacy hardware on machines might need upgrades – for MES to collect data automatically, machines may need sensors or communication modules that they previously lacked. Upgrading or retrofitting equipment for connectivity can be expensive or even infeasible on ancient machines, potentially limiting the scope of MES on such equipment. In some cases, companies mitigate this by using IoT gateway devices to capture signals non-intrusively, but this adds to the technical challenge. Compatibility issues with legacy technology can therefore slow the roll-out of MES or force the use of hybrid solutions (where parts of the process remain manual). A careful audit of the existing automation landscape, along with a roadmap for what needs updating, is a wise step before implementing a MES.

Transient Disruptions and Learning Curve: During the initial phase of MES adoption, companies may experience a temporary decline in productivity as employees adapt to the new system. There is a learning curve for operators to become proficient with MES terminals or handheld devices, as well as for IT staff to fine-tune system performance. It's not uncommon to encounter unexpected issues early on – for example, network overloads due to all machines sending data, or system errors that halt production because the MES is enforcing a rule too strictly. These teething problems can disrupt production if not managed, which is why many firms do pilot runs or phased rollouts (one line or area at a time) to stabilise the system. Managing expectations is essential: the full benefits of MES often accrue over months; in the short term, there will be challenges to overcome.

Despite these challenges, most companies that persist with MES implementation find that the long-term benefits outweigh the difficulties. The key is mitigation and planning: acknowledging these limitations and tackling them with clear strategies (integration plans, training programs, phased deployments, etc.)[103]. Additionally, support from leadership in providing necessary resources and championing the changes will significantly influence the success of MES adoption. In conclusion, MES projects are not trivial – they represent a significant change in how a factory operates – but with proper change management, technical planning, and realistic expectations, the hurdles can be overcome, paving the way for the numerous benefits discussed earlier.

Real-World Examples and Case Studies

To illustrate the role of MES in practice, it's helpful to examine how various industries implement MES to address their unique manufacturing challenges. MES solutions are employed across a wide range of sectors – from highly automated automotive plants to regulated pharmaceutical production, each with distinct drivers and results. Below are examples and case studies from several sectors, highlighting how MES is applied and the benefits achieved:

Automotive Manufacturing

The automotive industry was one of the early adopters of MES technology, driven by the need for high throughput, consistent quality, and traceability in vehicle production. Modern automotive plants are highly complex, assembling thousands of parts into vehicles with model variations and optional features – managing this complexity is impossible without robust digital systems like MES.

Use of MES: In an automotive assembly line, MES coordinates the entire production sequence, from body shop welding to final assembly. It ensures that the correct variant and options are built into each vehicle (often by reading a VIN or build sheet and orchestrating line-side supplies accordingly) and tracks each car through the hundreds of stations in the plant. MES provides real-time production status, enabling tight scheduling and synchronisation of parallel operations (for example, making sure that sub-assemblies like engines or seats are delivered Just-in-Time to be installed in the correct car). It also collects data at every station – including torque values from fastening tools, results of quality checks (such as leak tests or wheel alignment measurements), and more- to build a complete digital birth history of each vehicle. This level of control and data is crucial for automotive quality and safety standards.

Benefits: Automotive manufacturers report that MES helps improve efficiency, quality, and traceability across the factory floor[105]. By bridging the gap between the ERP system and shop-floor operations, MES provides real-time data and process optimisation that reduces waste and ensures production schedules are met [105]. For example, if a minor stoppage occurs in one zone of the assembly line, MES can adjust the downstream pace or reroute tasks to prevent a major line halt. Automotive MES implementations often result in reduced rework because issues are caught at the point of occurrence (no car leaves a station without passing the MES-driven quality checks). Traceability is another key benefit – MES enables automakers to trace every component (down to a batch of steel or a lot of airbags) that went into a specific vehicle, which is invaluable if a defect is discovered later. Large recalls can be avoided or minimised by quickly identifying the scope of affected vehicles through MES genealogy records. Additionally, MES helps with regulatory compliance (e.g., ensuring each car undergoes specific safety tests and has them documented) and provides data for continuous improvement in the plant (such as identifying that a particular model's assembly is slower, prompting process improvements).

MES in Action: A case study with a major automotive manufacturer illustrated these benefits: The company implemented an MES to unify production monitoring across multiple plants. As a result, they gained improved visibility and streamlined workflows, allowing them to adapt quickly to changes in production requirements while maintaining cost control and precision[105]. Another example is Daimler Trucks North America, which integrated MES with SCADA using a platform that gave both operators and management access to real-time production data. This integration allowed Daimler to adapt to production changes more easily and improve their traceability and efficiency, ultimately ensuring the reliable delivery of trucks with the desired configurations[97]. These cases demonstrate that MES has become essential in automotive manufacturing to meet the industry's high standards for volume, variability, and quality.

Aerospace and Defence Manufacturing

Aerospace manufacturing involves complex, long-cycle production with a heavy emphasis on quality and regulatory compliance. Aircraft components and systems must meet strict standards (FAA regulations, military specifications, etc.), and the manufacturing process is often highly customised and documentation-intensive. For these reasons,

MES has become a cornerstone in aerospace factories and maintenance/repair operations.

Use of MES: In aerospace, MES manages processes such as component fabrication, composite layup, machining, and the final assembly of large structures (like fuselages or wings), as well as the integration of systems (such as avionics and engines). A single aircraft can take weeks or months to assemble, and MES provides the tools to track progress and coordinate tasks among various teams. It enforces process steps (e.g., ensuring all torque checks on an airframe are signed off in sequence), and it often integrates with quality assurance systems to capture inspection results, non-conformance reports, and rework actions. MES in aerospace is also used to manage tooling and calibration, as many tools and machines require certification to tight tolerances. The MES ensures that only calibrated tools are used and automatically logs when a tool's calibration is due. Crucially, MES serves as the system of record for the as-built configuration of aircraft, linking the serial numbers of parts and subassemblies to each tail number of an aeroplane and maintaining the genealogy required for airworthiness certifications.

Benefits: The aerospace sector reaps significant benefits from MES in terms of quality control, compliance, and paperless documentation. By replacing paper travellers and building books with electronic records, MES helps aerospace manufacturers go largely paperless, which not only saves time but also reduces errors associated with manual record-keeping. For example, GE Aerospace implemented an MES (GE Proficy) in one of its facilities and reported moving from a very paper-heavy process to an automated one, which made build instructions and drawings easier to manage and significantly improved their ability to track work-in-progress across different cells[106]. They were previously struggling with siloed cells and a lack of visibility, and after MES, they had better control of factory floor operations and automated delivery of real-time work instructions to operators, reducing confusion and delays[106].

Compliance is another key benefit. Aerospace companies must maintain meticulous production records for each aircraft or component (for audit and certification). MES ensures that the digital thread of these records is intact. Every assembly step and inspection can be reviewed, and electronic signatures capture who performed each task, as required by regulations. This dramatically simplifies preparing for audits by aviation authorities or demonstrating compliance with standards such as AS9100 (quality management for the aerospace industry). Additionally, the analytics from MES (such as defect Pareto charts and throughput data) help aerospace manufacturers drive continuous improvement in a context where even small efficiency gains can result in significant savings, given the high cost of aircraft production. Overall, MES in aerospace yields a more controlled, transparent, and efficient build process, which translates to improved delivery schedules and right-first-time quality–critical factors when building complex aerospace products.

MES in Action: BAE Systems (a defence and aerospace manufacturer) implemented an MES across their operations to replace legacy systems and paper processes. According to a case study, this MES deployment put people first by simplifying user interfaces and workflows, resulting in higher user adoption. The outcome was a more agile production capable of handling design changes and complex assembly with fewer errors, as well as enhanced compliance reporting (with every action traceable) – a vital capability in the defence sector. Likewise, Airbus and Boeing have invested in MES (and broader MOM systems) to manage their global production and supplier networks, ensuring that components built worldwide meet the required specifications and are correctly recorded. These real-world adoptions show that in aerospace, MES is not just a productivity tool but a necessary backbone for quality and compliance.

Pharmaceutical and Life Sciences

The pharmaceutical industry has some of the most stringent manufacturing requirements, driven by Good Manufacturing Practices (GMP) and regulatory bodies such as the FDA. Manufacturing of drugs (whether small-molecule pharmaceuticals or biopharmaceuticals) and medical devices demands precise processes, complete traceability, and thorough documentation (batch records, electronic signatures, etc.).

MES has become a critical technology in the pharmaceutical industry, enabling what is often referred to as Pharma 4.0 —the digitisation of drug manufacturing.

Use of MES: In a pharmaceutical production setting, MES often functions as the Electronic Batch Record (EBR) system. Traditional paper batch records – massive binders capturing every step in producing a batch of product – are replaced by an MES that guides operators through each step of a batch process (weighing, mixing, reaction, filling, etc.) and captures data automatically from equipment (such as temperatures, pH, machine setpoints) as well as manual inputs (like confirming an ingredient addition or recording a visual inspection). The MES will enforce process adherence: for example, it won't let the batch proceed to the next step until a required quality sample is recorded or a supervisor electronically signs off on a deviation. It also manages recipes and calculations, ensuring that if a batch size is adjusted, all ingredient quantities and process parameters are correctly scaled. In pharmaceutical packaging operations, MES tracks serialised products (for traceability in line with regulations for drug serialisation) and manages in-line quality checks (like vision systems checking packaging or label printing) with immediate data collection.

Benefits: The primary benefits of MES in the pharmaceutical industry are compliance, quality assurance, and improved efficiency in batch record management. By using MES as an EBR, companies drastically reduce the possibility of errors that arise from manual recording. For instance, MES can automatically verify that the correct lot of raw material is being used by scanning barcodes, which might be misrecorded on paper. This ensures right-first-time manufacturing, critical when a mistake could mean scrapping an entire batch of expensive product. A case study from a pharmaceutical plant in Japan (Sakamoto Yakuhin Kogyo) that implemented a pharma-specific MES (Werum PAS-X) highlights these benefits: they achieved a GMP-compliant, entirely paperless operation with automatic collection of manufacturing data and integration to equipment and ERP, allowing production even to be started with a simple tablet interface[107]. The result was not only increased production efficiency but also easier compliance audits, as regulators could be given access to a complete, untampered electronic batch record that demonstrated adherence to all protocols [107].

Another benefit is reduced batch release times. Traditionally, after a batch is produced, a quality assurance team spends days reviewing the paper batch record for completeness and correctness before releasing the product. With MES, review-by-exception is possible – since the data is captured in real time and validated, QA can often release the batch much faster, as any deviations or alarms are already noted in the system. This means products reach the market sooner, or manufacturing throughput is higher with the same assets. Additionally, MES facilitates audit readiness: during an FDA audit, having electronic records that can be queried and displayed instantly (with a complete history of who did what and when) is a significant advantage, often making audits proceed more smoothly.

MES also supports Pharma 4.0 initiatives, such as continuous manufacturing (where drugs are produced in a constant process rather than in batches), by providing the control and data infrastructure needed for real-time quality monitoring and control. In such advanced setups, MES may integrate with PAT (Process Analytical Technology) instruments and act on their data to adjust process parameters in real-time, ensuring quality remains within specifications – essentially merging MES with real-time control for quality.

MES in Action: A global pharmaceutical company implemented MES across multiple formulation and packaging sites. As a result, they reported that batch record review time was cut by over 50%, and they experienced fewer deviations because the MES prevented many common errors (like missed process steps or data transcription mistakes). In one plant, the switch to MES and the ensuing paperless operation resulted in an estimated reduction of tens of thousands of paper pages per year, saving countless hours of documentation effort. Another case is NKP Pharma (as noted in a Rockwell Automation case study), where FactoryTalk Pharma MES was used to achieve compliance and deliver advanced data for decision-making [108]. The MES enabled NKP to integrate quality checks tightly, consistently meeting regulatory standards while also reducing manufacturing costs. These examples underscore that for pharmaceutical manufacturers, MES is often a linchpin technology that ensures product quality, patient safety, and regulatory compliance, while also improving operational efficiency.

Food and Beverage Manufacturing

Food and Beverage (F&B) manufacturers operate in a competitive environment characterised by thin margins, high throughput, and stringent food safety requirements. They must manage everything from recipe consistency to allergen segregation and traceability for ingredients sourced globally. MES in F&B is increasingly utilised to maintain quality and safety, maximise equipment utilisation (as many F&B processes operate 24/7), and handle frequent changeovers for various products or flavours.

Use of MES: In a food processing or beverage plant, an MES might manage processes such as mixing, cooking, bottling, and packaging. It will track each batch of product from the intake of raw ingredients through processing and into finished goods. Key functionalities include recipe management (ensuring that the exact recipe parameters are followed for each batch), traceability (tracking lot codes of ingredients through to final products and recording processing conditions), and performance monitoring (OEE monitoring on production lines to identify downtime causes, etc.). MES can integrate with sensors and PLCs on equipment such as pasteurizers, ovens, or fillers to capture critical control point data – including temperatures, pressures, and fill weights – which are essential for both quality and regulatory compliance (e.g., HACCP requirements). It also often connects to weighing scales and scanning systems to validate ingredients, preventing, for example, an allergenic ingredient from being added to a product that shouldn't contain it. In packaging, MES helps manage coding/labelling (correct date codes, lot codes) and can link with vision systems or checkweighers to ensure packaging integrity.

Benefits: For F&B manufacturers, traceability and compliance are the top benefits of MES. Food safety regulations (and good practice) demand the ability to trace ingredients from farm to fork. MES records every step of production, creating an electronic lot genealogy that links raw ingredient lots to finished product batches. This traceability is crucial for handling recalls efficiently. Suppose a particular ingredient is found to be contaminated or a specific day's production has an issue. In that case, MES enables the rapid identification of which products are affected, so only those specific lots are recalled, protecting consumers and minimising business impact. Moreover, MES helps ensure consistent product quality by enforcing standard processes and monitoring critical quality parameters. With real-time data tracking from raw material intake to final packaging, any deviation (like a temperature dropping below a safe pasteurisation level or a mixing time that's too short) can trigger an immediate response (such as halting the line and quarantining the affected batch)[101]. This reduces the risk of substandard products reaching the market, thereby protecting the brand's reputation.

MES also delivers efficiency gains. Many F&B plants use MES for OEE improvement – by analysing downtime and speed loss reasons on lines (due to stops for cleaning, changeovers, maintenance, supply issues), they can optimise scheduling and maintenance to maximise runtime[101]. For example, MES might reveal that a bottling line experiences frequent micro-stops due to label jams; addressing this issue could significantly increase output. Additionally, MES can coordinate production scheduling with cleaning cycles or product changeovers to minimise waste (like scheduling allergen-free products in sequence to avoid extra cleaning). Real-time visibility into processes enables supervisors to respond to issues quickly, thereby reducing prolonged downtime and waste generation.

Another benefit is material and inventory management. MES provides real-time inventory of WIP and semi-finished goods, helping avoid situations where a production line starves for materials because someone forgot to replenish an ingredient – the MES can alert when an ingredient bin is running low, for instance. It can also reduce overproduction and inventory holding by aligning production closely with demand signals, in conjunction with an ERP system.

MES in Action: AriZona Beverages, a large beverage producer, deployed an MES integrated with SCADA and their SAP ERP at a 600,000-square-foot facility. This MES integration gave management, quality, and maintenance teams access to real-time production data across the plant via their corporate network[97]. The results included improved visibility and coordination, as they could immediately see if a filler machine stopped or if a batch was out of specification and take action. Production data flowed automatically into SAP for inventory and batch tracking. Another example is a dairy products company that used MES to implement electronic traceability. By scanning raw milk lot codes at reception and then tracking processing through pasteurisation to packaging, they could trace any single carton of milk back to the farm and the collection time. When a contamination issue occurred with a particular farm's milk, they were able to isolate all products made from that lot within hours, demonstrating the power of MES-enabled traceability. On the efficiency side, a case study (Baldwin Richardson Foods, a sauce manufacturer) reported that MES helped operators track downtime and analyse performance in real time, resulting in a measurable increase in OEE on their production lines[109]. These cases demonstrate how MES enables F&B manufacturers to consistently meet quality and safety standards while driving operational improvements in throughput and waste reduction.

Prominent MES Platforms and Vendors

The MES market encompasses a diverse range of software providers, including industrial automation giants, enterprise software companies, and specialised vendors. When selecting an MES, manufacturers often evaluate vendors based on industry fit, functionality, integration capabilities, and support. Some of the prominent MES platforms and vendors in the market today include:

Siemens Opcenter Execution: Siemens Digital Industries Software offers the Opcenter suite (formerly known as SIMATIC IT and Camstar in parts), which includes MES solutions for discrete and process industries. Siemens Opcenter Execution is recognised for its comprehensive functionality and robust integration with automation and

PLM (Siemens also offers the MOM integration framework, based on ISA-95). It has a strong presence in electronics, semiconductor, medical device, and CPG industries, among others. Siemens has been noted as a leader in the MES space, particularly among vendors coming from a PLM background[110].

Rockwell Automation FactoryTalk MES (and Plex): Rockwell Automation provides MES solutions under its FactoryTalk suite, such as FactoryTalk ProductionCentre (often used in life sciences and automotive)[111]. Rockwell's MES is deeply integrated with their control and SCADA offerings (given Rockwell's PLC/HMI dominance), which is advantageous for integration on the shop floor. In recent years, Rockwell acquired Plex Systems, a cloud-native MES/ERP platform now offered as Plex MES, targeting automotive suppliers and general manufacturing specifically, with a software-as-a-service model. This combination of traditional on-premise MES and cloud MES gives Rockwell a broad portfolio.

GE Digital Proficy (Plant Applications): GE Digital (now part of GE Vernova) has the Proficy Smart Factory MES, historically known as Proficy Plant Applications. GE's MES has strong roots in high-volume manufacturing and has been widely used in industries such as automotive, aerospace, and FMCG, including within GE's factories. It offers modules for production execution, quality, and batch processes, and is known for its robust real-time data handling (leveraging GE's expertise in industrial data historians). GE's MES also often emphasises analytics and IIoT integration, aligning with GE's broader digital transformation vision.

AVEVA (Schneider Electric) MES (formerly Wonderware MES): Wonderware was a pioneer in industrial software, and its MES solution (now under AVEVA, after Schneider Electric's acquisition and merger with AVEVA) remains a popular choice. AVEVA MES (formerly Wonderware) is used in a variety of industries, particularly food & beverage, chemicals, and other batch/process manufacturing, as well as packaging operations[112]. It's known for a strong user interface and integration with SCADA (Wonderware's InTouch HMI is widely used).

Schneider Electric/AVEVA's solution often appeals to those already using AVEVA for SCADA and plant monitoring, providing a seamless extension to MES capabilities. Recent updates focus on multi-site MES and standardised KPI reporting across enterprises[113].

Dassault Systèmes DELMIA Apriso: DELMIA Apriso (from Dassault Systèmes) is an MES/MOM platform that is part of the broader DELMIA portfolio (which also covers digital manufacturing and planning). Apriso MES is recognised for its flexibility and has been successfully implemented in various industries, including automotive, aerospace, and industrial equipment. Given Dassault's PLM heritage (CATIA, ENOVIA), Apriso MES is often chosen by companies looking to tightly integrate design (PLM) with manufacturing execution as part of a "digital thread". It supports global manufacturing operations with features that enable standardised processes across plants.

Honeywell Manufacturing Execution System: Honeywell offers MES solutions, particularly geared towards the process industries (like oil & gas, petrochemicals, and chemicals) as well as pharmaceuticals. As an automation and controls company, Honeywell's MES (and the broader Honeywell Connected Plant portfolio) integrates with DCS/PLC systems, often emphasising batch execution and electronic logbooks for operators. Honeywell's offerings may be tailored to highly regulated or continuous process environments.

ABB Ability Manufacturing Operations Management: ABB's Ability platform encompasses MES/MOM solutions, primarily targeting industries such as mining, minerals, pharmaceuticals, and speciality chemicals. ABB's MES typically leverages their strength in control systems and electrification – it's often used when a tight integration with ABB control systems is desired. ABB's approach sometimes blends MES with advanced control and manufacturing optimisation tools.

SAP Manufacturing Execution (ME/MII): SAP, known for ERP, also provides MES capabilities. SAP ME (Manufacturing Execution) and SAP MII (Manufacturing Integration and Intelligence) are solutions that integrate directly with SAP ERP. SAP ME (originating from the acquisition of Visiprise years ago) is used in discrete industries, such as high-tech and aerospace. In contrast, SAP MII serves as a bridge for integrating shop-floor data with SAP systems (often used for building custom dashboards or connections to non-SAP shop-floor devices). Companies that are heavily invested in SAP's ecosystem might opt for SAP's MES to have a more unified IT landscape. SAP's approach often emphasises the seamless flow of data from MES to ERP (such as the automatic posting of production confirmations and goods movements).

Oracle Manufacturing and MES Solutions: Oracle offers MES functionalities within its Oracle Manufacturing Cloud and has previously provided Oracle MES for Process Manufacturing. While Oracle is more well-known in the ERP and database domains, some manufacturers using Oracle ERP have leveraged Oracle's MES modules or partners that offer MES solutions integrated with Oracle's platform.

PTC ThingWorx and MES Partnerships: PTC, through its ThingWorx IoT platform, has entered the realm of smart manufacturing. While PTC doesn't market a traditional MES, it partners with companies (and has acquired one, leading to collaboration with Rockwell) to offer integrated solutions that combine IoT, augmented reality, and MES functionality. PTC's focus is on connectivity (ThingWorx IIoT) and the "digital twin" concept, but by connecting to MES or embedding execution capabilities, they provide a modern take on operations management.

Other Specialized MES Vendors: There are many other notable MES providers: Critical Manufacturing (by ASM) which targets electronics and semiconductor manufacturing with a modern MES; 42Q which offers a cloud MES (initially spun out of Sanmina corp. for electronics manufacturing); iBASEt Solumina which specializes in aerospace, defense, and complex industrial manufacturing MES/MOM; Sepasoft MES which is built on the Inductive Automation Ignition platform (often used for MES integrated with Ignition SCADA in mid-size operations); Aptean (which has MES for process industries like food/beverage); MPDV (a Germany-based MES provider with the HYDRA MES, widely used in Europe); and others. The landscape includes both broad, platform-type MES and industry-specific solutions.

Leading MES vendors typically offer comprehensive functionality, but each has strengths in specific industries or integration scenarios. For instance, Siemens, Rockwell, and AVEVA (Wonderware) are often cited for their strengths in automation integration; Dassault and PTC for PLM integration; SAP and Oracle for ERP integration; and specialised vendors for deep industry-specific features. Many of these vendors have been recognised in analyst reports (like the Gartner Magic Quadrant or IDC MarketScape) as leaders in the MES space[110].

When evaluating MES platforms, manufacturers consider not just the feature set, but also factors like ease of use, scalability (can it handle multiple plants? global operations?), support for emerging technologies (cloud, mobility, AI), and the vendor's roadmap. The good news is that the MES market is mature, so there are proven solutions for virtually every manufacturing scenario – the challenge is choosing the right one and implementing it effectively.

Emerging Trends in MES: Smart Manufacturing and Industry 4.0

Manufacturing Execution Systems are continually evolving, especially as part of the broader trend toward Smart Manufacturing or Industry 4.0. In recent years, several key trends have emerged that are shaping the future of MES:

Cloud-Based MES and Multi-Site Management: Traditionally, MES systems were deployed on-premises at each factory site. However, with the rise of cloud computing, there is a clear trend toward Cloud MES solutions. Cloud-based MES (offered in a SaaS model or hosted in corporate data centres/private clouds) can provide a unified platform for multiple production sites, enabling enterprise-wide visibility and easier roll-out of standard processes. Early adopters are using cloud technology to break down silos and get a global view of production operations in real time, rather than each plant operating in its own MES bubble[100]. A cloud MES can aggregate data from many facilities, which helps in benchmarking performance across plants and allows corporate teams to drill down into any factory or line as needed[100]. It also simplifies software updates and maintenance (done centrally), and can reduce infrastructure costs per site. A critical enabler for this trend is improved security and reliability in cloud offerings, addressing concerns that manufacturers have historically had about depending on the cloud for mission-critical operations. We also see hybrid approaches – for example, on-premise MES for real-time control functions, but cloud for analytics and cross-site coordination. A practical example of cloud MES impact: a company with ten factories implemented a cloud MES and found they could introduce a process improvement developed at one site to all other sites within days by updating the central system configuration, whereas before it might have taken months to propagate changes through separate systems.

IIoT and Sensor Integration – The MES/IoT Convergence: Industry 4.0 emphasises the use of the Industrial Internet of Things (IIoT) – a network of smart sensors and connected devices throughout the factory. MES is converging with IoT in the sense that MES can act as a consumer and orchestrator of IoT data. Modern MES platforms are designed to ingest high volumes of time-series data from sensors (such as machine temperatures, vibration readings, and energy consumption) and integrate that data with context (orders, lots, and operators). This creates a richer data environment for decision-making. For example, by integrating with IoT data, an MES can correlate a slight increase in motor vibration (as detected by a sensor) with a specific production run and alert maintenance personnel or adjust the production plan to prevent a breakdown. Some MES vendors incorporate IoT connectivity standards (MQTT, OPC UA, etc.) natively to easily capture data from the myriad of devices on the shop floor. The trend is towards more granular and real-time data feeding the MES. This also leads to blurred lines between MES and other systems; sometimes, an IoT platform with analytics might assume roles traditionally performed by MES, or vice versa. The overarching goal is to have a fully instrumented and connected factory, where the MES serves as the central command system, providing visibility into the state of everything through IoT connectivity.

Artificial Intelligence (AI) and Machine Learning (ML) Integration: AI and ML are increasingly being applied in manufacturing, and MES is a natural platform to leverage these technologies, given the wealth of data it manages. AI/ML integration with MES can take several forms. One use case is predictive analytics, which involves using machine learning models on historical MES data, combined with sensor inputs, to predict outcomes such as machine failures (predictive maintenance) or quality issues. For instance, an AI model might analyse patterns in temperature and humidity data alongside production speed and previous defect data to predict when a batch is at risk of going out of specification. If the model flags a risk, the MES can trigger an intervention (adjust process parameters or perform an equipment check) before a defect occurs[96, 100]. Another application is advanced scheduling and optimisation – AI algorithms can optimise production schedules in ways that consider far more variables than a human planner could, potentially improving asset utilisation and minimising changeover time beyond what standard scheduling rules achieve. We also see AI aiding in visual quality inspection: MES can integrate with machine vision systems where ML models detect defects on products (e.g., identifying a surface flaw on a part or a misapplied label on packaging). When the AI model identifies an issue, it sends it to MES to record a defect and potentially divert the product for rework automatically [100]. In summary, AI/ML enhances MES by providing predictive insights and automated decision-making that complement the rule-based, deterministic logic of traditional MES. This trend is still maturing, but is expected to grow as manufacturers seek to make their operations more autonomous and data-driven. The convergence of AI and MES is sometimes referred to as "smart MES" or AI-enabled MES, indicating an MES that not only executes tasks but also learns and optimises from data.

Mobility and MES Apps: The modern workforce expects mobile and user-friendly interfaces, and MES is adapting to meet these expectations. Instead of operators being tied to fixed terminals, many MES now offer mobile apps or web interfaces that can run on tablets, smartphones, or wearable devices on the shop floor. This means supervisors can receive alerts and check production status on a tablet as they walk the floor, or an operator can scan a barcode with a handheld device to log a material consumption. Mobile MES interfaces can increase productivity by delivering information at the point of use and enabling quick data entry without requiring users to walk back and forth to a central station. Additionally, having MES data accessible on mobile devices extends visibility to maintenance crews or managers who might not be at a desk. Data accessibility from any device is becoming a standard expectation[99]. Some systems also integrate Augmented Reality (AR) on tablets or smart glasses. For example, an operator using an AR app could see real-time MES data or work instructions overlaid on the actual equipment they are looking at, which can reduce errors and training time[96]. Overall, improved user experience through modern UI design and mobility is a notable trend, helping drive higher MES adoption on the shop floor.

MES as Part of a Digital Twin/Digital Thread: In the context of Industry 4.0, there's much talk of the Digital Twin (a digital replica of a physical asset or process) and the Digital Thread (the connected data flow across the product lifecycle). MES plays a critical role in these concepts. A digital twin of a production line, for instance, relies on real-time data from MES and control systems to simulate and predict performance. We see MES integrating with simulation and digital modelling tools so that there's a tight link between the virtual model and the real operation. For example, a digital twin might simulate how a change in a machine setting could increase output; MES provides the baseline data and would enact the change if approved. The digital thread refers to the continuity of data from design to manufacturing to service – MES contributes to the manufacturing portion of that thread, documenting how a product was made. Emerging MES solutions thus emphasise open data architectures and APIs that allow MES data to feed into enterprise analytics, product lifecycle analytics, and even field performance data (in cases of connected products). The trend is toward a more integrated information landscape, where MES is not a standalone system but an integral part of a larger smart manufacturing ecosystem that also involves PLM, ERP, supply chain systems, and IoT platforms [99].

Microservices and Modular MES Architectures: Traditional MES software can be quite monolithic. A new trend is the design of MES as a collection of microservices or modular applications. This aligns with industry movement toward more agile software development and deployment. Instead of a single heavy MES application, vendors are breaking MES functionality into smaller services (e.g., a service for production scheduling, a service for quality management, etc.) that can be independently deployed and scaled. This microservice approach often goes hand in hand with cloud deployment and containerization. The benefit for manufacturers is greater flexibility – they can adopt pieces of functionality incrementally and potentially mix and match services. It also allows for easier updates (one service can be updated without affecting the entire system) and can improve reliability (if one microservice has an issue, it may not crash the entire MES). Some leading vendors have begun to re-architect their MES offerings based on these principles, marketing them as "MES platforms" that are highly extensible. For example, an MES might expose RESTful APIs or event streams that can be subscribed to by third-party or custom applications, enabling an ecosystem of apps around the core MES.

MES and Edge Computing: With IoT and massive data streams, another trend is running specific MES or MOM functions on the edge (closer to the machines). Edge computing can filter and process data from machines locally for fast response – for instance, performing quick analysis or storing data during network outages, then syncing with the central MES. Some solutions are embedding mini-MES capabilities in edge devices, which is helpful for remote sites or when connectivity to the central system is limited. This way, critical production can continue with local decision support even if the cloud or central server is temporarily unreachable.

Greater Emphasis on Interoperability and Standards: To realise these trends, interoperability is key. The adoption of standards such as ISA-95 (for enterprise-to-shop floor data models) and OPC UA (for device-level data communication) is becoming increasingly important. We also see the rise of industry-specific data standards that MES may support (for example, in electronics manufacturing, the IPC-CFX standard for machine connectivity). In addition, organisations like MESA International continue to work on defining new MES best practices in the era of Industry 4.0, and groups such as ISO/IEC have begun standardising aspects of smart manufacturing architectures. All of this is pushing MES vendors to ensure their systems can communicate easily with others, avoiding proprietary lock-in. Manufacturers now often demand that MES have proven interfaces to their existing ERP, PLM, or automation systems (for instance, templates to connect to SAP or Oracle, or certified OPC connectivity).

In combination, these emerging trends indicate that MES is transforming from a plant-centric execution engine to a more connected, intelligent orchestration system that operates across the enterprise and leverages advanced technologies. The MES of the future is expected to be more flexible (to adapt quickly to new product introductions or changes), more intelligent (using AI to assist decisions), and more connected (both vertically to business systems and horizontally to machines and supply chain). This evolution positions MES as a key enabler of the smart factory vision, where manufacturing is highly automated, adaptable, and optimised end-to-end through digital technologies[99, 114].

Standards and Frameworks for MES (ISA-95 and Others)

The field of manufacturing execution is supported by several essential standards and frameworks that provide common definitions, models, and guidelines for effective execution. These standards help ensure that MES implementations are structured and that they can integrate more easily with other systems. The most notable framework related to MES is ISA-95.

ISA-95 (IEC/ISO 62264) – Enterprise-Control System Integration: ISA-95 is an international standard (originally released by the International Society of Automation and also published as IEC 62264) that defines the interface between enterprise business systems and control systems. It provides a formal model of manufacturing activities organised into hierarchical levels, which has been widely adopted in the industry[96, 104]. The standard's hierarchy is often depicted as the automation pyramid (shown earlier in the figure), summarised as:

- Level 4: Business Planning and Logistics (typically ERP systems and other enterprise applications).

- Level 3: Manufacturing Operations Management (the realm of MES/MOM systems).

- Level 2: Monitoring, Supervisory Control (SCADA, HMI systems that coordinate multiple control loops).

- Level 1: Control (PLC, DCS – direct machine control).

- Level 0: The physical process itself (sensors and actuators acting on the process)[104, 115].

In ISA-95, MES resides at Level 3, acting as a buffer and translator between the business level and the control level [95, 96]. ISA-95 not only defines these levels but also specifies standard terminology and data models for the information exchanged between levels. For example, ISA-95 models define what a "production schedule" contains, how "production performance" can be summarised, what a "material lot" is, and so on. By standardising terminology and data structures, ISA-95 makes it easier for different systems (from various vendors) to communicate without misunderstanding[96]. This reduces integration errors and effort – an MES built with ISA-95 in mind will have data objects for production orders, operations, equipment, personnel, etc., that correspond to how an ERP like SAP might send that information, since SAP also aligns with ISA-95 models for its manufacturing integrations[96].

ISA-95 comprises multiple parts, covering models for activities, object models, and formal methods for implementing interfaces. One of the key concepts from ISA-95 is the division of functions into production, quality, maintenance, and inventory operations within Level 3. It essentially laid the groundwork for what functions MES/MOM should encompass (which aligned well with the earlier MESA model). Companies use ISA-95 in various ways: as a guideline for MES requirements (ensuring they consider all necessary functions), as a reference for developing or selecting MES solutions (checking if the solution supports ISA-95 models), and for designing integration interfaces (for example, using ISA-95 XML/B2MML schemas to link MES with ERP). Many vendors will advertise their compliance with ISA-95 or support for ISA-95 standards, indicating that their software fits into this layered model cleanly and can communicate as expected.

In practical terms, adopting ISA-95 can enable a more plug-and-play integration. For instance, if an ERP sends a work order to MES, ISA-95 defines the content (like material requirements, operations, schedule, etc.) so that the MES knows how to interpret it[104]. Without such standards, every ERP-MES pair would need a custom mapping. ISA-95 has thus been fundamental in reducing the effort required to integrate enterprise and control systems, which is one reason it's highly valued in multi-vendor environments[104].

MESA International and the MESA Model: Before ISA-95 gained prominence, the Manufacturing Enterprise Solutions Association (MESA) played a key role in defining MES. MESA is an industry association that published the MESA-11 model in 1997, identifying 11 core MES functions (which we listed earlier)[95, 96]. While not a formal standard, the MESA model became a de facto framework for understanding the scope of MES. Even today, MES vendors refer to those core functions when describing their offerings. MESA continues to provide guidance, whitepapers, and best practices for MES/MOM and the value it brings to manufacturing. MESA's work complements ISA-95: MESA focused on MES functionality, whereas ISA-95 focuses on integration and hierarchy. The two are often referenced together, and indeed, ISA-95's models incorporated the MESA functions when defining the activities of Level 3[95].

Batch Control (ISA-88) and MES: Another relevant standard, especially for process industries (chemicals, pharma, food), is ISA-88 (also known as IEC 61512), which defines models and terminology for batch control. ISA-88 is more focused on the control recipe level (Level 2) – it describes how to structure recipes, equipment hierarchies, and batch records. However, MES for batch processes (often called Batch Execution Systems when focusing on recipe execution) is directly influenced by ISA-88. Many MES systems that handle batch production (like recipes for a drug or formula for a beverage) use ISA-88 concepts. They have a recipe manager, they manage batch IDs, and they interface with batch control systems (like PLC or DCS that execute the recipe phases). If the MES includes batch dispatching or EBR, it likely aligns with ISA-88 so that it can properly interface with control systems following that standard. For example, a pharma MES might generate an ISA-88 compliant batch record that can be understood by any auditor familiar with ISA-88 structures.

Quality and Data Standards: Although not specifically related to MES, standards such as ISO 9001 (Quality Management Systems) impact MES, as one of its key roles is to help enforce quality procedures. Additionally, sector-specific GMP guidelines (Good Manufacturing Practices) in the pharmaceutical industry, or GAMP (Good Automated Manufacturing Practice) for the validation of automated systems, provide frameworks that MES must adhere to in regulated industries. For instance, GAMP 5 provides guidelines on validating MES software to ensure it meets regulatory requirements. Therefore, in a pharmaceutical context, following GAMP is a standard procedure when implementing MES. MES often also must comply with 21 CFR Part 11 (for electronic signatures and records in FDA-regulated industries). So while Part 11 is a regulation, not a framework, it essentially mandates specific MES capabilities (secure user login, electronic signature with meaning of signature, audit trails that are immutable). Most MES destined for pharma/biotech have built-in features to be Part 11 compliant.

Newer Frameworks: With the advent of Industry 4.0, new frameworks and reference architectures have emerged. For example, the Industrial Internet Reference Architecture (IIRA) or the RAMI 4.0 (Reference Architecture Model Industry 4.0) from Germany. These are high-level frameworks that integrate all aspects of Industry 4.0, and MES would be a component within them (for instance, RAMI 4.0 has a three-dimensional grid that includes hierarchy levels similar to ISA-95, life cycle value streams, and layers of IT representation). These frameworks aim to ensure that as factories digitalise, all systems (including MES) fit into a cohesive architecture. While these are still evolving, they reiterate the importance of standards for interoperability. For instance, RAMI 4.0 works closely with OPC UA and emerging standards for asset administration shells, which could be relevant to how MES identifies and communicates with equipment.

In practice, ISA-95 remains the cornerstone standard that any MES professional should understand. It's commonly used for structuring discussions about integration. For example, when mapping out how an MES project will connect to ERP, teams often break down the data exchanges in ISA-95 terms (like defining "Level 4 to Level 3 transactions" for production orders and "Level 3 to Level 4 transactions" for performance results). It provides a common language for vendors, integrators, and users[96]. As noted by SAP, using ISA-95 consistent models "reduces the risk of error when integrating manufacturing sites with business systems". This is a powerful endorsement of how standards make MES projects smoother.

To conclude, standards and frameworks such as ISA-95 (IEC 62264) and the MESA model have been instrumental in guiding the development and implementation of MES. They help define what MES should do and how it should integrate, which in turn protects manufacturers' investments by ensuring their MES can communicate in a larger ecosystem. As MES continues to evolve with Industry 4.0, adherence to standards and open architectures will become even more critical – it will enable the MES to act as the connective tissue in smart factories, interoperating with a wide array of devices and systems in a standardised manner. Industry professionals should leverage these frameworks for planning MES implementations and insist on standards compliance when evaluating MES solutions to ensure future-proof, scalable deployments[96, 104].

PLM in Manufacturing: Driving Innovation Across the Product Lifecycle

Defining PLM and Its Role in Digital Manufacturing

Product Lifecycle Management (PLM) is a strategic approach and software solution for managing a product's lifecycle end-to-end – from initial concept and design, through manufacturing and support, all the way to service and disposal[116]. In essence, PLM integrates people, data, processes, and business systems, serving as an information backbone that connects everyone involved in a product's development. By centralising all product data and automating workflows, PLM ensures that teams across engineering, manufacturing, procurement, and service are working with the latest information and coordinated processes[116, 117].

In the context of digital manufacturing, PLM plays a pivotal role as the hub of the digital thread – the connected data flow that links every stage of the product lifecycle. Modern factories rely on a multitude of digital tools (CAD for design, simulation software for analysis, ERP for resource planning, MES for execution on the shop floor, etc.). PLM serves as the connective tissue between these tools, ensuring that changes made in design flow through to manufacturing processes and supply chain plans seamlessly. It provides a single source of truth for product definitions (parts, bills of materials, specifications, etc.), enabling digital manufacturing initiatives like automation, simulation, and real-time feedback loops. In short, PLM is the backbone that supports Industry 4.0 by unifying the digital models, data, and processes needed to design and produce complex products in a coordinated, efficient manner[118, 119].

By standardising and organising all product information, PLM enables companies in digital manufacturing to innovate faster and with greater confidence. Teams can collaborate globally on 3D designs, run simulations on virtual prototypes, plan manufacturing workflows, and track quality metrics – all within an integrated environment. This ensures that as products move from the digital realm (CAD models, simulations) to the physical realm (production and deployment), there is full traceability and alignment. As a result, PLM is essential for streamlining product development in modern manufacturing, reducing errors due to miscommunication, and enabling techniques like model-based systems engineering and digital twins (virtual representations of physical products) to flourish[118, 120].

Core PLM Functionalities

A modern PLM system comprises a wide range of functionalities that together manage all aspects of product data and processes. Some of the core PLM functionalities include Bill of Materials management, change and configuration management, document control, and product data management. These are described below:

Bill of Materials (BOM) Management

At the heart of any product definition is the Bill of Materials (BOM) – a structured list of components and sub-assemblies needed to build the product. PLM provides intelligent BOM management capabilities to create and handle complex, multi-level BOMs throughout the product lifecycle[121]. Unlike manual spreadsheets or isolated systems, PLM's BOM management ensures that engineering BOMs (EBOMs) from design, manufacturing BOMs (MBOMs) for production, and even service BOMs remain linked and up-to-date. A well-managed BOM in PLM improves quality and cost control by clearly defining the product structure and linking each part to relevant data (CAD models, specifications, approved suppliers, etc.)[121].

Benefits of PLM-based BOM management include faster product release cycles (with easier BOM revisions and approvals), greater accuracy through structured relationships between parts, and better visibility via consolidated BOM reporting[121]. For example, teams can quickly generate a "flattened" BOM to see all components needed, or filter a BOM for a specific configuration or effectivity. By serving as the central BOM repository, PLM makes sure that all departments (engineering, procurement, manufacturing, etc.) are referencing the same, current BOM information. This reduces errors (like someone ordering the wrong part revision) and prevents the costly mistakes that can occur when BOM data is scattered in silos. BOM management in PLM also extends to handling approved manufacturer lists (AML) and approved supplier lists (ASL), giving design engineers instant visibility into which suppliers and parts are qualified for use[121]. This holistic control of the BOM through PLM is critical in industries with complex products (e.g. automotive or electronics), as it keeps the entire organisation aligned on what is being built at every stage.

Engineering Change Management

Change is inevitable during product development – designs get refined, issues are found, and requirements shift. Engineering Change Management (ECM) in PLM defines a formal process to propose, review, approve, and implement changes to the product[121]. This usually involves workflows for Engineering Change Requests (ECRs) and Engineering Change Orders (ECOs). A PLM system provides a controlled environment where change proposals can be evaluated with all relevant data in context – designs, BOM, cost, impact on other systems – and then approved changes are systematically applied so that every stakeholder knows exactly what changed and why.

With PLM-driven change management, organisations minimise errors and rework by ensuring changes are documented, reviewed by the right people, and traced to requirements or issues[121]. Automated workflows route change requests to responsible parties and track status, keeping change cycles as short as possible[121]. By maintaining a full history of changes (what was changed, who approved it, when it was effective), PLM supports compliance and audit requirements, which is vital in regulated industries. Furthermore, PLM can automatically update linked information when a change is released (for example, notifying manufacturing to use a new part revision or updating the BOM to a new configuration). In practice, robust change management means a change made by an engineer is efficiently propagated to procurement (so they buy the updated part) and to production (so they build to the latest design), avoiding costly scenarios where different departments act on outdated information. Formal change control is a cornerstone of quality: it helps ensure the final product meets specifications and safety standards by preventing unintended changes [121].

In summary, PLM's change management functionality brings discipline to product updates, preventing "engineering chaos". It keeps the development process on schedule and budget by evaluating change impacts beforehand[121] and enforcing approvals, and it provides a closed-loop system where design modifications are visible to all downstream stakeholders in real time.

Configuration Management

Closely related to change management is Configuration Management, which in the PLM context means maintaining consistency of a product's performance and functional attributes with its design and documentation throughout its life. In practice, configuration management ensures that at any point in the lifecycle, one can reconstruct the exact configuration of a product, including the specific versions of all components, software, and documentation. PLM systems enable this by tracking versions and revisions of parts and documents, managing product variants, and controlling the release of configurations for production or delivery.

For complex products (like aircraft, automobiles, or electronics), there may be multiple configurations and variants in production simultaneously (different models, options, or customer-specific versions). PLM allows creation of configuration rules or variant BOMs (sometimes called 150% BOM or modular BOM), which can be filtered to derive a specific configured BOM for a given product instance or variant. This dynamic BOM filtering and variant management is a key PLM capability, often integrated with configuration features in CAD and ERP[122]. By using PLM to manage configurations, companies can ensure that the "as-designed", "as-built", and "as-serviced" representations of a product are all aligned. For example, when a specific aircraft (by serial number) undergoes a repair, the service team can trace exactly which design configuration it was built to, and ensure replacement parts match that configuration.

In PLM, configuration management also means maintaining baseline releases – capturing a snapshot of the entire product structure and documentation at a release milestone. This is crucial for regulatory compliance and quality (it provides traceability: one can always find what design was approved and built at a given time). Standards such as ISO 10007 (guidelines for configuration management) emphasise having proper identification and control of configurations, which PLM systems support by uniquely identifying each item and its approved versions. Additionally, PLM often integrates with software configuration management or application lifecycle management (ALM) tools for products that include software, ensuring hardware and software changes are synchronised. Overall, PLM-based configuration management gives organisations the ability to navigate complexity – they can confidently manage hundreds of product variants and changes, knowing that each delivered unit's configuration is documented and can be reproduced or analysed if needed.

Document Control (Product Documentation Management)

Another core functionality of PLM is document management and control for all product-related documents. Modern products generate a massive volume of documents – CAD drawings, requirements specifications, test reports, work instructions, regulatory certificates, user manuals, and more. PLM provides a centralised repository where all these documents are stored with proper version control and access control[121]. Instead of having files scattered on network drives or individual computers, PLM ensures that documents (often referred to as product records or technical data packages) are organised by product and revision, and are easily retrievable by authorised users.

Crucially, PLM enforces revision control and audit trails on documents. Teams can't accidentally use an obsolete procedure or drawing because the PLM system will clearly indicate the latest approved version and can even watermark prints as "uncontrolled" if not accessed from the system[121]. This is indispensable for maintaining standards like ISO

9001, which require rigorous control of documents and records. For instance, quality policies, Standard Operating Procedures (SOPs), assembly instructions, and design history files (DHF) can all be managed in PLM with appropriate reviews and approval workflows[121].

By using PLM for document control, organisations gain benefits like: centralised access (everyone sees the same up-to-date documents in one place), improved collaboration (multiple team members or even suppliers can review and comment on a document within the PLM environment), and elimination of redundant or lost files[121]. It also saves time – instead of hunting through emails or folders for the latest spec sheet, a user can search the PLM system and trust the results. Document control is often tied into other PLM processes: for example, releasing a change in PLM might automatically update a related PDF in the system and notify users. All these capabilities help reduce mistakes and delays. In regulated industries (like medical devices or aerospace), this is absolutely critical because auditors will want to see that every requirement and design output is documented and that changes were controlled; PLM provides that backbone of traceability.

Product Data Management (PDM)

Product Data Management (PDM) is sometimes considered a subset or the foundation of PLM. It focuses on managing the detailed design data, primarily for engineering teams – things like CAD models, drawings, part files, and their version history[117]. Early PLM systems evolved out of PDM systems that were created to handle the challenge of large CAD files and version control[116]. Key PDM capabilities include secure storage of design files, check-in/check-out for collaborative editing, revision tracking, and metadata management (so that parts and files can be catalogued and searched).

While PLM covers the entire lifecycle (concept to end-of-life across multiple departments), PDM is centred on the design and development phase. However, an effective PLM implementation almost always includes strong PDM functionality at its core[117]. For example, Siemens Teamcenter or PTC Windchill will have modules to integrate with CAD tools (SolidWorks, NX, Creo, etc.) so that when designers save their work, the CAD files and associated BOM data are captured in the PLM's database with the proper relationships and versioning. PDM ensures that an engineer in one location can't accidentally overwrite another's design, and that the company's intellectual property (IP) (its CAD and engineering data) is backed up and managed systematically.

PDM also typically manages the engineering Bill of Materials and ties parts to drawings – so one can click a part in PLM/PDM and find its CAD model and drawing, or vice versa[117]. It handles access control too: for instance, only authorised personnel can see or edit certain designs (necessary for IT security of sensitive product info). As companies grow, they often start with PDM for engineers and then expand to PLM across the enterprise[117]. PDM helps eliminate the chaos of disorganised design files, thereby laying the groundwork for broader PLM processes. In summary, PDM is a crucial PLM function that tames the complexity of engineering data and ensures the design content is properly governed – it feeds the rest of the PLM system with the authoritative product definitions created by engineering.

Other key PLM functionalities: In addition to the above, PLM systems often encompass other modules such as requirements management (capturing and linking customer or regulatory requirements to the design data), project and program management (tracking project timelines, stage-gates, and deliverables within the PLM environment[121]), quality management (managing FMEAs, CAPAs, test results), and supplier collaboration portals (to involve vendors in design reviews or RFQs[121]). All these pieces integrate to form a comprehensive platform. Ultimately, the core components of PLM work in concert to unify people, processes, and technology, providing a single digital platform to manage products efficiently from inception to retirement[117].

Integration of PLM with Other Digital Manufacturing Systems (MES, ERP, CAD, Simulation)

One of the most powerful aspects of PLM is how it integrates with other enterprise systems to enable a seamless digital manufacturing ecosystem. In a typical manufacturing company, PLM does not exist in isolation – it must connect to CAD tools, ERP (Enterprise Resource

Planning), MES (Manufacturing Execution Systems), and often various simulation and analysis tools. Integration ensures that data flows automatically between design, planning, and execution stages, supporting the broader concept of a digital thread across the product lifecycle.

- PLM and CAD/Simulation Integration: PLM is inherently linked to CAD, since it stores and manages the CAD models and drawings produced during design. Most PLM systems offer direct integrations (plug-ins) for major CAD software, so designers can save files into the PLM repository without leaving their CAD environment[117]. This integration enables features like part numbering, revision control from within CAD, and the ability to run impact analyses (e.g., "where-used" searches to see what assemblies a part is used in). Additionally, PLM integration with simulation tools (CAE – Computer-Aided Engineering) means that simulation models, results, and validation reports can be associated with the design data. For instance, an FEA (finite element analysis) result file can be linked to the specific CAD version of the part it analysed. Some modern PLM platforms (like Dassault's 3DEXPERIENCE or Siemens' integrated suite) provide unified environments where CAD, CAM, and CAE all connect to the PLM backbone[120]. This allows digital manufacturing scenarios such as virtual prototyping and digital twins: Engineers can simulate the product's performance in a virtual environment and feed insights (like a needed design change) back into the PLM system, which then triggers an engineering change workflow. Overall, tight PLM-CAD integration ensures that the authoritative design data in PLM is always linked to the latest geometry and simulations, enabling concurrent engineering and reducing manual data transfer.

- PLM and ERP Integration: PLM and ERP are complementary systems often described as addressing different domains – PLM is focused on product innovation and design data, whereas ERP handles resource planning, procurement, finance, and the execution of manufacturing and business processes[102]. Yet, they share critical touchpoints. A typical integration point is the BOM: the PLM (as design system) may own the engineering BOM, while the ERP system needs a manufacturing BOM and inventory of parts to plan purchasing and production. Integrating PLM with ERP ensures that when engineering releases a new part or BOM in PLM, that information flows into ERP so that materials can be ordered and production can be scheduled[102]. Similarly, changes released in PLM (ECOs) can update item master data or BOMs in ERP automatically, preventing discrepancies between what engineering designed and what manufacturing builds. Another integration aspect is feeding cost and supplier data from ERP back into PLM, so that designers have visibility into cost implications and supplier status when making decisions. When PLM-ERP integration is done effectively under a clear IT strategy, redundant data entry is eliminated, and each system does what it's best at, yielding a combined value greater than either alone[102]. The result is better coordination between design and the business execution: for example, as soon as a product design is finalised in PLM, ERP is ready to generate purchase orders for components and arrange production slots, compressing the time to market.

- PLM and MES Integration: Manufacturing Execution Systems (MES) control and monitor the production floor, tracking work orders, operations, machine data, and quality checks in real-time. Integrating PLM with MES establishes a closed-loop manufacturing process. Through this integration, the "as-designed" product definition (from PLM) is communicated to the shop floor via MES. Concretely, PLM can provide MES with information like manufacturing process plans, work instructions, and product configurations for building. In return, MES can feed production results and quality data back into PLM. For example, MES might report that for a specific serial number of a product, a particular component was substituted or a deviation occurred; PLM can capture that as part of the product record (useful for traceability and future design improvements). Real-time feedback from MES allows engineers to see how their designs are performing in production and to address manufacturability issues quickly[123]. Moreover, PLM-MES integration helps enforce build consistency: the MES will guide operators to use the correct drawings and specs (fetched from PLM) for each job, which reduces errors and rework. If an engineering change happens, the updated instructions and drawings are delivered to MES promptly so the factory builds to the new requirements. In essence, this integration contributes to a digital thread from design to production, ensuring the product is built exactly as intended and capturing any discrepancies or shop floor improvements back into the PLM knowledge base[102].

- Integration with Other Tools (e.g., CRM, SCM, etc.): In digital manufacturing, PLM may also integrate with Customer Relationship Management (CRM) systems or supply chain management tools. For instance, requirements or issues logged by customers (in a CRM or service system) can be linked to PLM as new requirements or change requests, thus feeding real-world input into the product development loop. Integration with supply chain systems or supplier portals allows external partners to receive the latest specs directly or to submit part changes into the PLM workflow securely[121]. Additionally, with the rise of IoT (Internet of Things), PLM platforms are beginning to integrate with IoT data platforms to incorporate field data into the product record, forming a loop where usage data informs new design iterations (for example, PTC's Windchill integrates with their ThingWorx IoT platform for this purpose).

Benefits of Integration: When PLM, ERP, MES, and other systems are well-integrated, companies realise significant benefits: streamlined workflows with less manual data transcription, elimination of costly errors due to inconsistent data, and faster decision-making thanks to end-to-end visibility[102]. For example, integration helps eliminate rework, one of the most expensive consequences of poor coordination. If design changes don't reach manufacturing in time, it can result in building the wrong parts (waste) or needing product rework – integration mitigates that by synchronising data across systems[102]. It also enhances agility:

a change in market demand might be captured in PLM as a new variant, passed to ERP to adjust production planning, and then to MES to adjust shop schedules, all in a coordinated flow. Additionally, a well-integrated PLM/ERP/MES landscape strengthens data security by reducing the need for ad-hoc spreadsheets and emails that can leak sensitive information. Product data stays within managed systems with proper access controls[102]. In summary, integration of PLM with other digital manufacturing systems is a foundation for Industry 4.0's connected enterprise, allowing the digital thread to run through design, planning, and production and enabling a truly responsive manufacturing operation[102].

Benefits of PLM in Digital Manufacturing

Implementing PLM provides numerous benefits that directly improve business outcomes for manufacturing organisations. Key advantages include better product quality, faster time-to-market, enhanced collaboration (across teams and geographies), and the enablement of a continuous digital thread of information. These benefits translate into tangible metrics – for example, companies using PLM have reported up to 30% faster time-to-market and 25% lower development costs, according to industry benchmarks[124]. Below, we detail these major benefits:

Improved Product Quality and Compliance

PLM contributes to higher product quality by enforcing repeatable processes and providing traceability throughout the product's development. Because all product data (requirements, designs, validations, changes, etc.) is managed in one place, it's easier to spot inconsistencies and ensure that the final product meets its specifications and regulatory requirements. Quality is built in by design – PLM systems often integrate with Quality Management processes (like tracking of non-conformances, test results, and corrective actions). For instance, PLM can ensure that a design cannot be released without required validation tests being attached and approved, or it can trigger quality checks in manufacturing when certain high-risk parts are being built.

Traceability in PLM means that if a defect is found (say, a faulty component), one can quickly trace which product versions and serial numbers used that component and take action (facilitating efficient root-cause analysis and recalls if needed). Moreover, PLM's rigorous change control prevents unauthorised or ad-hoc changes that could degrade quality. Every change is evaluated and verified, reducing the chance of mistakes slipping through. Regulatory compliance is also bolstered: industries like aerospace, automotive, medical devices, etc., must comply with standards (FAA regulations, FDA 21 CFR Part 820, ISO standards like ISO 9001) that demand thorough documentation and control. PLM systems inherently support these by maintaining the Design History File, Device Master Record, or other compliance documentation as part of the product record, and by providing the audit trails of who approved what and when[118]. In short, PLM makes quality processes part of the development DNA, leading to more reliable products and easier compliance reporting (e.g., generating an ISO 9001 compliance report or technical file for CE marking from PLM data).

An emerging aspect of quality is closing the loop with field data: PLM integrated with IoT can collect performance data from products in use and feed that to engineers. This enables a proactive quality improvement approach – often termed Quality by Design (QbD) – where lessons from actual product performance drive design enhancements[118]. By leveraging the digital thread, PLM ensures that such data doesn't live in isolation but is used to refine product quality over successive iterations continually.

Reduced Time-to-Market

Accelerating the development cycle to launch products faster is a critical competitive edge in many industries. PLM helps reduce time-to-market in several ways. First, by streamlining workflows and automating routine tasks, PLM cuts down on administrative overhead (for example, automatically notifying approvers and tracking their input, rather than engineers chasing signatures manually)[124]. This means decisions and approvals happen faster. Second, PLM's central data repository minimises the time wasted searching for information or waiting for someone to send the latest file. Everyone can access what they need in real-time, enabling parallel work streams. For instance, as soon as a design is released in PLM, manufacturing engineers can begin process planning off that data – the design and manufacturing plan activities can overlap rather than proceed sequentially.

Collaboration tools in PLM (discussed more below) also significantly speed up development by breaking down barriers between departments; issues get resolved quicker when the entire team shares visibility. Additionally, PLM enables re-use of existing designs and knowledge, which is a huge time saver. Engineers can search the PLM system to find if a similar part or solution already exists instead of reinventing the wheel. This can shrink design cycles drastically. In industries like electronics or consumer goods, PLM supports rapid iteration and introduction of new models by managing product line variants efficiently – configuring new product versions from a base of existing components is much faster than creating each from scratch.

A concrete example of PLM's impact: Volvo Construction Equipment found that using PLM as the backbone for an industrialised digital thread improved cross-team efficiency and accelerated time-to-market for new products[125]. Similarly, automotive companies leverage PLM to bring high-quality vehicles to market faster to meet demand[126]. By integrating the efforts of design, simulation, and manufacturing planning early through PLM, companies can compress the overall timeline and respond swiftly to market changes or new customer requirements. The result is often a shorter development cycle, sometimes cutting months off schedules, which allows firms to capture market opportunities and revenue earlier than competitors.

Enhanced Collaboration Across Teams and Geographies

In today's globalised engineering environment, teams are often distributed across different locations and disciplines. PLM greatly enhances collaboration by providing a common platform and data source that all stakeholders access. Instead of each team (mechanical, electrical, software, manufacturing, suppliers, etc.) working in isolation with their data, PLM connects them. Everyone from design engineers to procurement to field service can see relevant product information and status in the same system[127]. This single source of truth eliminates the miscommunications that arise from fragmented data, such as one department working off an outdated drawing.

Collaboration features in PLM might include real-time co-authoring of data, discussion threads or commenting on design items, and automated notifications. For example, when an engineer uploads a new design iteration, the manufacturing engineer subscribed to that part can be instantly alerted to review manufacturability. Teams across geographies can work virtually together on a product structure or a document within PLM, avoiding endless email chains. By breaking down silos, PLM ensures that knowledge flows freely: a change made by engineering is visible to procurement to adjust supplier plans, or a concern raised by a factory technician can be logged into PLM for engineers to consider.

Additionally, PLM extends collaboration to external partners in a controlled way. Suppliers or customers can be given access to specific parts of the PLM data (for instance, a contract manufacturer might access the BOM and drawings for the product they build, and can provide feedback or receive updates via the PLM supplier portal[121]). This kind of extended enterprise collaboration reduces delays and errors in hand-offs.

Collaboration is not just about communication, but about making decisions collaboratively with full context. With PLM's dashboards and analytics, cross-functional meetings can be more productive because everyone is looking at the same live data (like project status or a KPI dashboard of open issues). The enhanced collaboration leads to benefits like fewer design revisions (since manufacturing and service input can be incorporated early), improved innovation (ideas from various teams are captured), and a more agile organisation overall. PLM is often seen as an enabler for concurrent engineering – multiple disciplines working in parallel, as opposed to the old sequential "over-the-wall" approach. Companies that implement PLM frequently report that it strengthens teamwork and aligns the organisation toward common goals, since all departments are literally on the same page regarding product information[(124)].

Enabling a Digital Thread and Digital Twin Strategy

Perhaps one of the most transformative benefits of PLM in the era of digital manufacturing is enabling the digital thread and digital twin strategies. The digital thread is the idea of having a continuous thread of data that connects every phase of the product lifecycle, creating an integrated view of the product from design to end-of-life. PLM is the foundation of this thread[(118)]. By managing all the data and changes in one place and integrating with other systems as described, PLM makes it possible to trace any requirement or change forward and backwards across the lifecycle. For example, a single requirement can be traced to design parts, to manufacturing processes, to test results, and to field performance. This traceability is the digital thread in action, and it provides unprecedented visibility and insight. It allows organisations to leverage data from one stage (say, manufacturing yield data from MES or failure data from field service) to inform decisions at another stage (like design improvements or materials changes).

Hand-in-hand with the digital thread is the concept of the digital twin – a digital replica of a physical product or system that can be used for simulation, monitoring, and predictive analysis. PLM plays a crucial role in digital twin implementations by housing the master definition of the product. The digital twin of a product unit (e.g., a specific car or a jet engine) can be maintained by taking the master design from PLM and then linking it with instance-specific data (like sensor readings or usage history). PLM ensures that the digital twin remains accurate to the physical product by feeding it any changes or updates that occurred during manufacturing or maintenance. Conversely, data from the operational twin (through IoT sensors) can loop back into PLM to inform new designs, completing a feedback loop.

Modern PLM solutions (referred to as PLM 4.0) explicitly support establishing and monitoring the digital thread that connects the "voices" of various domains – machines (IoT data), products (design/Digital Twin data), factories (production data), and customers (feedback) – into a cohesive whole[119]. By doing so, companies achieve faster innovation cycles, better decision-making and improved quality because insights are drawn from the full spectrum of lifecycle data rather than siloed snapshots[119]. For example, aerospace firms use digital threads to track each aircraft through design, production, and service, so they can predict maintenance needs or analyse the impact of a design change on past and future fleet performance. This wouldn't be feasible without a PLM backbone to organise and provide context for all that data. In manufacturing, a digital thread enabled by PLM means that if a problem is discovered in a product already shipped, one can quickly find all related data (materials used, who approved the design, what other products might have the issue, etc.) in one chain of information, vastly reducing the response time and ensuring nothing is overlooked.

In summary, PLM's enablement of digital thread and twins leads to data-driven decision making and a more proactive approach to product lifecycle management. It connects traditionally separate activities (design, test, production, support) into a continuous loop, thereby helping organisations to not only make better products faster, but also manage them intelligently throughout their life in the field.

Key Challenges in PLM Implementation

While PLM offers substantial benefits, implementing a PLM system in an organisation is not without challenges. Industry professionals often encounter hurdles such as data silos and migration issues, resistance to change among users, integration complexity with existing tools, and concerns about scalability and performance as the system grows. Understanding these challenges is crucial to planning a successful PLM adoption:

Breaking Down Existing Data Silos: Many companies adopting PLM are moving from a situation where product data is scattered – some in CAD files on local drives, some in ERP, some in spreadsheets. Consolidating this into a single PLM repository is a big task. One challenge is data migration – ensuring all legacy data (and its history) is brought into PLM accurately and consistently. Often, data from different sources has inconsistencies or lacks standardisation (e.g., multiple naming conventions for the same part), which can impede the effective use of PLM[128]. If not addressed, companies might create a new silo (the PLM system) that doesn't truly talk to others. To overcome this, a thorough data cleansing and standardisation effort is needed upfront. PLM vendors provide tools for import, but the strategy must define what data to migrate (not every historical detail may need to be moved) and how to map it. Failure in this area can lead to garbage in, garbage out – a PLM that users don't trust because the data is incomplete or incorrect. Therefore, establishing a single source of truth may require significant preparatory work.

User Adoption and Resistance to Change: Implementing PLM often entails new processes and workflows, which means people have to alter their daily work processes. Resistance to change is a common human factor challenge[128]. Engineers who are used to their filing systems, or project managers accustomed to using email and spreadsheets, might initially see PLM as extra overhead or fear losing control. Low user adoption can undermine even a technically successful PLM deployment[128]. To mitigate this, change management is essential: organisations should communicate the "why" and benefits of PLM, involve end-users in the design/configuration of the system, and provide comprehensive training. Executive sponsorship plays a role here – when leadership actively advocates and requires the use of the PLM for all product data, it helps drive adoption. Gamifying adoption (celebrating early wins, showcasing improvements) can also turn sceptics into advocates. The key is to shift the culture from siloed thinking to a collaborative mindset, and that doesn't happen overnight. Companies that ignore the cultural shift aspect risk the PLM system becoming underutilised, with employees finding workarounds that perpetuate silos (like continuing side databases or personal spreadsheets).

Integration Complexity: A PLM system must integrate with numerous other systems (CAD, ERP, MES, etc.) as discussed earlier. Technically, setting up and maintaining these integrations can be complex. Each interface might involve different tools, data mappings, and protocols (e.g., using middleware or API calls). If not done correctly, poor integration can result in data not flowing properly, creating new silos or data inconsistencies[128]. For example, if PLM-ERP integration fails, engineering might release a part in PLM, but purchasing doesn't see it in ERP, causing supply chain delays. Integration challenges also include synchronising different update cycles (a change in one system must trigger an update in another reliably) and handling errors or conflicts between systems. Furthermore, upgrades to any one system (say a new version of the ERP or CAD software) can break the integration unless carefully managed. For many companies, leveraging standard integration solutions or middleware (provided by PLM vendors or third parties) is helpful, but often some customisation is needed to fit specific business rules. It's critical to allocate enough time and resources in the implementation plan for integration testing and refinement. Ignoring this can lead to a disjointed system where PLM fails to deliver on its connectivity promise, and users experience issues such as mismatched part numbers between systems. Best practice suggests addressing integration in the early phases of implementation and not as an afterthought – consider PLM as part of an enterprise architecture strategy.

Scalability and Performance: As the volume of product data and the number of users grow, the PLM system must scale accordingly. Companies worry whether the chosen PLM can handle, say, a million parts in the database, or hundreds of concurrent users across the globe. An inflexible or unscalable PLM solution can become a bottleneck, hindering the organisation's growth or forcing expensive reimplementation later[128]. Cloud-based or modern PLM (PLM 4.0) solutions have made strides in scalability (with SaaS offerings that can elastically handle load). Still, legacy PLM deployments might struggle if not architected for it. Scalability also refers to functional scalability – as business needs evolve, can the PLM expand to cover new processes or integrate new tools? If a system is too rigid, adding a new product line or merging data from an acquisition could be very challenging. Organisations should thus assess not just current needs but also future requirements. Piloting the PLM with a smaller scope is common, but one must ensure the architecture (server sizing, database, network, etc.) is designed with headroom. Another aspect is maintaining performance – if opening a BOM or running a search in PLM takes too long, users will get frustrated. Regular performance tuning and possibly archiving of old data (or using data lake solutions for ancient big data) might be needed to keep the system responsive. Essentially, scalability planning is a challenge that must be met with careful infrastructure design and by choosing a PLM solution known to handle the company's industry scale (for instance, some PLM systems are used explicitly in enterprise contexts with massive datasets, while others target mid-size companies).

In addition to these, other common challenges include cost and ROI justification (PLM can be a significant investment, and leadership will want to ensure returns, which might require robust ROI models), vendor selection (choosing a PLM platform that fits the industry and integrates well; each of the leading PLM systems has strengths and weaknesses), and data security concerns (consolidating crown-jewel IP in one system means that system must be very secure). Many organisations follow best practices like phased implementation (not trying to "boil the ocean" all at once, but rolling out PLM capabilities in stages) and engaging experienced implementation partners to mitigate these challenges. Awareness and proactive management of these challenges are vital. For example, addressing resistance to change via a strong training program, or tackling data silo issues by involving IT and business analysts early to map out data flows. When done correctly, the effort pays off in a smoother PLM rollout and earlier realisation of benefits.

In summary, implementing PLM requires not just a technological change but a process and culture change. Overcoming data silos, getting user buy-in, ensuring integrations, and planning for growth are all part of the journey. Companies that navigate these challenges by setting clear goals, securing executive support, and following best practices (some of which are outlined in the last section on standards and best practices) set themselves up to fully leverage PLM's capabilities.

Real-World Use Cases and Industry Examples

PLM is employed across a variety of industries – from automobiles and aeroplanes to smartphones and consumer gadgets. Examining real-world use cases provides insight into how PLM drives success in different sectors:

Automotive Industry

The automotive industry was an early adopter of PLM due to the complexity of vehicle design and the need to coordinate large engineering teams and suppliers. Automakers use PLM to manage the entire lifecycle of a vehicle, ensuring that from the concept stage through design, production, and maintenance, all data is connected. By leveraging PLM, automotive companies can accelerate time-to-market for new models, ensure regulatory compliance, optimise resources globally, and enhance product quality[129].

For example, consider the development of a new car model. Thousands of components (mechanical parts, electronic controllers, software modules) need to be designed and integrated. A PLM system (like Siemens Teamcenter or Dassault Systèmes ENOVIA) will host all the CAD designs for these components, manage the BOMs for various configurations (different trims or region-specific variants), and keep track of changes as testing reveals issues or improvements. The collaboration aspect is crucial – design centres in other countries can work concurrently on the vehicle, and suppliers can be given access to the PLM to upload their part data or see the latest specifications for the components they supply. This was seen, for instance, in the development of electric vehicles, where the integration of new technologies (battery systems, autonomous driving tech) required close collaboration between software engineers, electrical engineers, and traditional automotive mechanical teams. PLM was the platform that brought all these together.

One well-known success story, Volvo Group, implemented a PLM-driven approach and reported that integrating their processes on a PLM backbone improved cross-team efficiency and allowed them to launch products faster[125]. Another example is how Ford Motor Company used

PLM tools to standardise their global engineering processes, enabling design reuse across regions and reducing its vehicle development time. Moreover, automotive companies must comply with strict safety standards (like crash regulations) and environmental regulations (emissions, end-of-life recycling directives). PLM systems ensure that all compliance documentation (e.g., material composition for REACH or ELV directives, test reports for safety) is attached to the vehicle's digital thread, making it easier to demonstrate compliance and trace any issue to its source.

PLM also supports automotive trends like mass customisation. Modern PLM configuration management allows a car company to manage a vast number of possible configurations (colour, engine type, features) under one product umbrella. This is sometimes called managing the "explosion of variants." A robust PLM setup can handle this by modular design management and rules (so-called product configurators within PLM). For manufacturing, PLM provides the MBOM and process plans to plants worldwide, ensuring consistency while allowing local adjustments. The benefit is not only speed but also reduced cost – reusing proven designs and processes across multiple car programs via PLM avoids duplication of effort.

Aerospace and Defence Industry

Aerospace products (like commercial aircraft, fighter jets, or spacecraft) are among the most complex engineering feats, involving millions of parts and extremely rigorous safety and quality requirements. PLM is indispensable in aerospace for managing this complexity and maintaining configuration control over decades (aircraft can be in service for 30+ years). Companies like Boeing and Airbus rely on PLM to coordinate their global design and manufacturing teams and supply chain, achieving improved efficiency, cost reduction, and better quality.

For instance, Boeing's 787 Dreamliner program involved design work shared between Boeing and numerous international partners. A central PLM system (Boeing has used Dassault Systèmes solutions and others) was used to share 3D models and drawings across the extended enterprise so that a wing designed in one country could fit precisely with a fuselage section built in another. The PLM also managed the complex certification data. Every part and material on a plane must be certified to meet aviation authority standards, and the chain of approvals and tests for each part is captured through PLM. Boeing's use of PLM and a digital model helped them reduce physical prototyping, but also taught lessons: earlier Airbus projects (like the A380) famously encountered issues when different teams used different CAD versions, highlighting the importance of a unified PLM environment to avoid data translation issues that can be extremely costly[130]. Both Airbus and Boeing have since doubled down on PLM to ensure all partners work on the exact digital mock-up. Airbus has a comprehensive PLM strategy called "Digital Design, Manufacturing and Services (DDMS)" to integrate their processes, reflecting how critical PLM is to their future.

Defence contractors like Lockheed Martin or Northrop Grumman similarly use PLM to manage programs such as fighter jets or satellites. Configuration management is a prime concern; for example, if a plane is updated with new avionics, PLM ensures there's a clear record of which aircraft tail numbers have which configurations. Traceability is life-critical in aerospace: if a defect is found in one component, PLM helps identify every aircraft or engine with that component. This kind of recall or service bulletin management would be nearly impossible without a digital thread. PLM also aids in the after-market service for aerospace, maintaining digital twins of individual aircraft that log all maintenance actions and modifications, which the likes of airlines and MRO (maintenance, repair, overhaul) organisations can access, often through OEM-provided PLM portals.

Furthermore, in space exploration, organisations like NASA or SpaceX employ PLM to integrate engineering disciplines – rocket design involves structural, thermal, software, electrical systems, etc. A PLM system orchestrates all the data so that a change in one system triggers analysis in others. It also helps with program management: meeting launch deadlines by tracking progress in a single system and ensuring that every component arriving for assembly is the correct version.

In summary, aerospace and defence have some of the most fully developed PLM use cases: PLM helps these companies design right the first time (because mistakes are incredibly expensive or dangerous), maintain control over product configurations over long lifecycles, and comply with stringent regulatory oversight. The result is improved efficiency and product quality – for instance, one study notes that both Boeing and Airbus saw significant improvements in efficiency and error reduction by implementing PLM strategies across their organisations.

Consumer Electronics and High-Tech Industry

In the fast-paced world of consumer electronics (think smartphones, laptops, wearables) and high-tech products, PLM is used to juggle rapid product release cycles, complex supply chains, and strict cost targets. Companies in this space face intense pressure to launch innovative products frequently (often multiple times a year) while coordinating manufacturing across many suppliers (for chips, displays, etc.) and ensuring high quality to avoid warranty issues. PLM helps by providing a central hub to manage product definitions and changes very quickly and to communicate instantly with manufacturing partners.

For example, a smartphone manufacturer will use PLM to manage the phone's BOM (which can have thousands of components down to resistors and capacitors), the mechanical designs of the casing, the firmware requirements, and even packaging and labelling artwork – all in one system. When the next model is being developed, they'll copy the previous BOM into PLM and update parts, a process expedited by PLM's controlled changes. Given the short development cycles, PLM's ability to maintain organisation and avoid mistakes is key. If a last-minute component swap is needed (say a chip is unavailable and a substitute will be used), a change order in PLM can ensure that everyone (the factory, procurement, quality teams) gets that update immediately, preventing a scenario where the wrong part is assembled.

Consumer electronics companies also must adhere to regulations like RoHS (Restriction of Hazardous Substances) or WEEE (Waste Electrical and Electronic Equipment) in their products. PLM systems often integrate compliance checks, ensuring that the BOM doesn't contain disallowed substances, and storing certificates from suppliers. For instance, if a new law limits the use of a specific flame retardant in plastics, the PLM can flag any part containing it and assist the team in finding alternatives. This proactive compliance management through PLM saves these companies from legal issues and market delays.

Another challenge in high-tech is collaboration across hardware and software teams. Products like smart home devices or consumer gadgets involve embedded software. Leading PLM solutions now integrate with Application Lifecycle Management (ALM) tools or have modules to manage software versions, linking them to the hardware BOM. This means when hardware changes (e.g., new sensor), PLM reminds to update the software requirement accordingly, ensuring the product as a whole works when built.

Real-world examples: Seagate Technology, a data storage device manufacturer, established an enterprise digital thread on a PLM foundation to connect teams and data, which improved cross-team efficiency and product introduction speed[125]. GoPro, a well-known action camera maker, as it grew, adopted PLM (specifically a cloud PLM solution) to manage its hardware designs, streamline engineering change across its distributed teams, and coordinate with manufacturing in Asia; this helped eliminate time-zone delays and errors in its product release process[125]. Similarly, companies like Dell or HP have used PLM to unify their design and manufacturing, leading to better configuration management of the numerous product variants they offer and more on-time product launches.

In consumer electronics, margins can be thin, and timing is everything (missing a holiday season can be disastrous). PLM provides the infrastructure to manage fast development sprints with accuracy. It also feeds into after-sales: for example, if a batch of laptops has a faulty component, PLM allows quick identification of the affected batch and provides customer service and repair teams with instructions on how to fix it. Arena Solutions (now a PTC company) has case studies of various high-tech electronics startups and mid-size firms where implementing PLM early helped them scale their operations without chaos, ensuring that even as their engineering headcount doubled, everyone stayed coordinated on a single system.

Summary of cross-industry use cases: Across automotive, aerospace, electronics, as well as other industries like industrial machinery, consumer goods, and medical devices, PLM's role is consistently to bring order, visibility, and efficiency to product development. Companies large and small use it to collaborate better and deliver products the first time. Leaders like Siemens, Boeing, and Airbus attribute improved efficiency, cost savings, and quality improvements to their PLM strategies. For instance, one can note that neglecting PLM can lead to lost opportunities and shorter product lifecycles, while effectively using PLM has helped these companies maintain product success in the market. These real-world examples reinforce that PLM is not just theoretical software, but a practical enabler of complex product realization in the real economy.

Prominent PLM Vendors and Platforms

Over the years, several PLM software platforms have become prominent in the industry, each backed by major software vendors and adopted by leading companies. Here we will highlight four of the most prominent PLM solutions – Siemens Teamcenter, Dassault Systèmes ENOVIA, PTC Windchill, and Autodesk Fusion Lifecycle (Fusion 360 Manage) – along with their key characteristics and areas of strength:

PLM Platform	Vendor & Key Features
Siemens Teamcenter	Siemens Digital Industries Software – Teamcenter is one of the most widely used enterprise PLM systems globally[124]. It offers broad and deep capabilities across product design, simulation, document management, BOM and configuration management, and manufacturing process planning. Teamcenter is known for its strong CAD integrations (especially with NX and Solid Edge, Siemens' CAD tools, but also multi-CAD) and scalability in supporting

	extensive implementations (automotive and aerospace OEMs are common users). It provides both on-premises and cloud (Teamcenter X) deployment options, and is praised for its "unmatched depth, breadth and usability" in connecting teams, tools, and processes enterprise-wide[124]. Siemens also integrates Teamcenter with other Siemens products like Tecnomatix (for digital manufacturing), giving a comprehensive portfolio from design to production.
Dassault Systèmes ENOVIA	Dassault Systèmes – ENOVIA is the PLM component of the 3DEXPERIENCE platform, which unifies design (CATIA), engineering, simulation (SIMULIA), and other applications on a single data model. ENOVIA has its roots in managing complex aerospace and automotive programs and is known for robust configuration management and collaboration capabilities in a 3D context. On 3DEXPERIENCE, ENOVIA goes beyond traditional PLM by integrating design, engineering, simulation, and manufacturing data in one environment, thus supporting model-based systems engineering and allowing all stakeholders to collaborate on a unified platform[120]. This platform approach enhances real-time collaboration and lifecycle traceability. ENOVIA's strengths are in industries like aerospace, defence, automotive, and industrial equipment – domains that Dassault has long served. It supports comprehensive BOM management, change control, requirements management, and more, with a modern web-based interface.

PTC Windchill	PTC – Windchill is a leading PLM solution known for its rich out-of-the-box functionality and strong PDM core. It has been adopted in industries ranging from industrial equipment to electronics and aerospace. Windchill provides standard modules for BOM and change management, document control, configuration management, and has options for requirements and project management. A key differentiator for PTC is its focus on the digital thread and incorporation of emerging technologies: Windchill is often paired with PTC's IoT platform (ThingWorx) and AR solutions (Vuforia), positioning it well for Industry 4.0 initiatives. PTC emphasises Windchill's ability to realise value quickly with its comprehensive set of core applications and its role as an enterprise foundation for the digital thread (125) ptc.com. Windchill can be deployed on-premises or as Windchill+ (a cloud SaaS offering)[125]. PTC also acquired Arena (a cloud PLM for mid-market and electronics/med device companies) and has FlexPLM (for retail and footwear industries), making their portfolio quite broad[125]. Windchill's integration with CREO (PTC's CAD) is tight, but it also supports multiple CAD systems. It's known for high configurability and a large user base in discrete manufacturing.
Autodesk Fusion Lifecycle	Autodesk – Fusion Lifecycle (recently rebranded as Fusion 360 Manage) is Autodesk's cloud-based PLM offering. It provides PLM capabilities with the convenience of a SaaS model – accessible via browser, minimal IT overhead, and quick setup with out-of-the-box workflows[131]. Autodesk positions it as an "instant-on PLM" that

	is easily configurable to a company's needs without heavy consulting or customisation [131]. Fusion Lifecycle is particularly attractive to companies already using Autodesk design tools (like Inventor, AutoCAD, or Fusion 360 CAD) as it integrates seamlessly with those, but it also supports multi-CAD environments. Core features include BOM and change management, NPI (New Product Introduction) project tracking, quality workflows, and supplier collaboration, delivered in a modular way (companies can implement only the needed apps and expand later). While historically Autodesk PLM is popular with small-to-mid-sized manufacturers and those in consumer products or machinery, it has been steadily gaining enterprise features. The emphasis is on ease of use and quick ROI, making the PLM approachable even for organisations with smaller IT teams.

Table: Main PLM vendors

Each of these platforms has its ecosystem of modules and partner extensions. For instance, Teamcenter and Windchill might be found in the same company in different departments (some companies use Teamcenter for CAD data management and Windchill for a specific product line, though ideally, one PLM is used enterprise-wide). Selection often depends on the existing tool landscape (CATIA users lean to ENOVIA, CREO users to Windchill, etc.), industry solutions (e.g., if you need apparel PLM, you might consider FlexPLM or a specialised solution). All of them support the core PLM functionalities we discussed, but with different philosophies: Siemens and Dassault emphasise an integrated suite from design to manufacturing, PTC emphasises tying PLM to IoT/digital thread, and Autodesk emphasises accessibility and integration with design-to-make cloud tools.

It's also worth noting other PLM vendors: Oracle offers PLM as part of its Cloud SCM suite (Oracle Agile PLM was historically known, now Oracle PLM Cloud, often used in process industries and high-tech). Aras Innovator provides a flexible, open-source PLM platform popular for its customisation and subscription model. SAP has PLM capabilities embedded in its ERP for companies deeply invested in SAP. Each solution has found a niche, but the four above are frequently cited in industry shortlists.

PLM's Evolving Role in Industry 4.0 and Smart Manufacturing

As manufacturing undergoes digital transformation (often termed Industry 4.0), PLM's role is expanding and evolving to meet new requirements. In the Industry 4.0 paradigm – characterised by IoT (Internet of Things), cyber-physical systems, cloud computing, and AI – PLM becomes even more crucial as a data orchestrator and process enabler.

One significant evolution is PLM 4.0, the next generation of PLM software that aligns with digital transformation goals. Modern PLM (PLM 4.0) is typically cloud-based, making it accessible globally and easier to scale and update with new features (SaaS model). This aligns with Industry 4.0's need for agility – companies can quickly onboard new users (even remote) and collaborate across geographies in real-time, thanks to cloud PLM. Moreover, PLM 4.0 embraces the integration of IoT data and connectivity with smart machines. As production equipment and products themselves become sources of data (via sensors), PLM serves to connect those "voices" of data back to the product definition. For example, Oracle describes that with PLM 4.0, companies can establish a digital thread connecting the "voices" of the machine (IoT), product (digital twin models), factory (operational data), and customer (feedback from usage) across the enterprise[119]. By having this connectivity, PLM ensures the enormous streams of data generated in Industry 4.0 are contextualised to the product life stages and can be used for actionable insights.

The digital twin concept is a cornerstone of Industry 4.0. PLM provides the authoritative source for the digital twin's configuration. Smart manufacturing leverages digital twins of not only the product but also the production system (e.g., a digital twin of a robotic assembly line). PLM intersects here by managing the data about the production process (tools, NC programs, etc., sometimes called Manufacturing Process Management or MPM) and linking it to product data. This means one can simulate the manufacturing process in a virtual environment (a form of digital twin of the factory) using the same data that will later drive the real factory. The outcome is optimised processes and the ability to foresee issues before they occur. For instance, aerospace manufacturers use this to virtually test assembly sequences of an aircraft in PLM's manufacturing planning module integrated with 3D simulations, refining them to save time on the shop floor.

Another aspect is the integration of Artificial Intelligence (AI) and analytics into PLM. The future of PLM sees AI assisting in tasks like automatically validating compliance, suggesting design improvements from past project data, or predicting project risks. Siemens notes trends like embedding AI for decision support and using IoT for closed-loop feedback between real-world product usage and development[124]. We already see some PLM systems incorporating analytics dashboards that, for example, highlight parts with frequent issues or suppliers that cause delays. As Industry 4.0 progresses, PLM could proactively optimise workflows (an AI agent might scan all open change requests and prioritise them based on impact analysis). AI could also help manage the deluge of data from connected products – summarising field issues, categorising them, and even initiating changes in PLM automatically when specific triggers are met.

Industry 4.0 also emphasises interoperability and standards, which affect PLM. Initiatives for smart manufacturing call for standardised ways to exchange data (e.g., using formats like ISO 10303 STEP for 3D models, or newer standards like OPC UA for machine data). PLM systems are adapting by supporting these standards, ensuring they can plug into a smart factory with minimal custom integration. The concept of the digital thread we discussed is inherently about using standards to flow data (one example: STEP AP242 is a standard for a managed model-based 3D integration, which some PLMs use to exchange data between CAD, PLM, and other tools without losing product structure or PMI - product manufacturing information).

In smart manufacturing, flexibility and customisation are essential – PLM's evolving role includes facilitating mass customisation through advanced configuration management and modular product architectures. As products become more configurable (sometimes down to an "order of one"), PLM systems handle the complexity by allowing highly parametric BOMs and automating the generation of variant data. They are incorporating more configuration lifecycle management capabilities so that even as thousands of unique product instances are produced, each one's data is traceable in the digital thread.

Digital manufacturing also involves technologies like 3D printing (additive manufacturing). PLM is evolving to manage additive manufacturing data – for example, managing printable files, material recipes, and linking them to design revisions. A PLM might trigger a 3D print job through an MES interface and capture the results.

Finally, PLM is becoming more user-friendly and accessible to a broader audience (not just engineers). Industry 4.0 pushes democratisation of data – shop floor workers or field technicians might use a tablet to access PLM data (via simplified apps or AR interfaces) to get instructions or report issues. This extends the reach of PLM from core engineering teams to practically everyone involved in the product lifecycle, fostering more integration of human feedback into the digital processes.

In essence, PLM is transitioning from a back-end engineering system to a central digital platform for innovation that is at the heart of a smart manufacturing enterprise. Its support for digital twin/thread strategies means it ensures continuity and integrity of data, which is the lifeblood of Industry 4.0 operations. As technologies advance, PLM continues to incorporate them – whether it's cloud, AI, or AR/VR – to maintain its role as the orchestrator of the product lifecycle in a fully digital, intelligent factory environment.

Industry Standards and Best Practices in PLM Implementation

Implementing PLM in alignment with industry standards and following best practices ensures that organisations get the most value from their investment while maintaining compliance and quality. Several standards are relevant:

ISO 10303 (STEP): ISO 10303, commonly known as STEP (Standard for the Exchange of Product model data), is a neutral file format standard that enables the exchange of product data between different software systems. STEP is crucial in PLM environments for achieving interoperability, especially in heterogeneous system landscapes. For example, a company might use CATIA for design but have a supplier using Creo; STEP files allow 3D models to be shared through the PLM without loss of geometry or metadata. Many PLM systems support STEP AP242, which covers managed model-based 3D engineering data, ensuring that the product structure and annotations are preserved when moving data. Adopting STEP and related standards in a PLM strategy is a best practice to avoid vendor lock-in and facilitate collaboration with partners (who can't all be on the same CAD/PLM). It future-proofs the product data – decades later, those neutral files can still be interpreted even if the original software is obsolete. Standards like JT (for lightweight 3D visualisation) also often come into play with PLM for visualisation across the enterprise.

ISO 9001 (Quality Management Systems): ISO 9001 sets out criteria for a quality management system (QMS). While not specific to PLM, many of the requirements of ISO 9001 (document control, managing design and development processes, change control, traceability, corrective actions) can be supported or enforced via the PLM system. Best practice is to integrate PLM and QMS processes so that the PLM becomes a tool to help comply with ISO 9001. For instance, ISO 9001 emphasises having controlled documents and records for all processes – the PLM's document management ensures only the latest released procedures are accessible, fulfilling this requirement. It also requires a defined design and development control process – a PLM-driven workflow with phase-gates, approvals, verification, and validation steps can be built in accordance with ISO 9001 guidelines. Companies implementing PLM often map their PLM workflows to the ISO 9001 clauses (e.g., design inputs, design outputs, design review, etc.). By doing so, when they go for ISO 9001 certification, the PLM can readily produce evidence of compliance (audit trails, approvals, etc.). In sectors like medical devices, which need ISO 13485 (a specific QMS standard) or aerospace's AS9100, PLM similarly helps manage the design history and change records required for compliance. Regulatory bodies appreciate when companies have a robust PLM because it usually correlates with better control over processes.

Other Relevant Standards: Depending on the industry, there are additional standards that interplay with PLM. For example, ISO 26262 (functional safety for automotive) requires configuration management and impact analysis for safety-related parts – a PLM with configuration and requirements traceability helps achieve that. ISO 15288 (systems engineering standards) and INCOSE systems engineering practices can be supported by linking requirements to design in PLM. There are also standards for configuration management, such as the guidance in ISO 10007, which outlines configuration management principles; a PLM implementation following these principles will include identifying configuration items, systematically controlling changes, and recording configuration status – all things PLM is well-suited for. CMMI (Capability Maturity Model Integration) for development processes can also be advanced by using PLM to institutionalise processes.

Best Practices in PLM Implementation: Aligning to standards is one side; the other is following industry best practices gleaned from numerous PLM deployments. Here are some key best practices:

Strong Executive Sponsorship and Clear Goals: Ensure you have executive buy-in and clearly define the objectives of the PLM implementation (reduce time to market by X%, improve reuse, comply with X standard, etc.). This aligns the effort with business objectives and secures resources. Without it, PLM projects can stall. A best practice is to quantify expected ROI (e.g., savings from less rework, faster ECO cycle times) to justify the project.

Phased Implementation (Start Small, Scale Fast): Rather than attempting a "big bang" rollout of all PLM modules and to all departments at once, it's often effective to start with a pilot or a subset (for example, begin with CAD data management and change control for one product line). This allows ironing out issues and demonstrating quick wins. The pilot acts as a proof-of-concept that can be showcased to a broader organisation. Once success is proven, iterate and expand PLM functionality (add BOM management, supplier collaboration, etc.) and onboard more teams. Many companies first implement core PDM, then gradually extend to full PLM[124].

Business Process Alignment and Simplification: Don't just implement PLM to mirror broken or convoluted processes. It's a chance to re-engineer and simplify workflows. Adopt industry best practices embedded in the PLM templates when possible (for instance, using a standard phase-gate NPI process or a best-practice change approval board workflow). This often means convincing stakeholders to adjust their ways of working to fit a more streamlined process that PLM supports out-of-the-box, rather than over-customising the PLM to every legacy quirk. Over-customisation is a trap that can make upgrades hard; best practice is to use configuration (which is easier to maintain) instead of heavy custom coding, and stick close to standard PLM processes, which themselves are based on industry experience.

Data Governance and Cleanup: As mentioned in the challenges, invest in cleaning and organising data before or during PLM implementation. Establish clear definitions: what constitutes a part vs. a document vs. a CAD model in the system, what naming conventions and attributes will be used. Set up a governance team or steering committee that monitors data quality in PLM (for example, periodically auditing that users are following the process for metadata entry, or that there aren't duplicate part entries). This governance ensures the PLM remains the trusted source of information (single source of truth). It's also wise to load critical historical data that might be needed for reference, but not clutter the system with unnecessary data (a balance must be struck).

Integration and IT Strategy: Plan the integration points early and decide system ownership of data. Best practice is to define clearly, for example, "PLM is a system of record for part master and BOM; ERP is a system of record for pricing and inventory." Then integrate accordingly so that each system pulls what it needs. Use proven middleware or APIs, and involve both PLM and ERP/MES experts in design workshops. Ensuring the PLM fits into the broader IT landscape (with proper infrastructure, backups, disaster recovery) is part of best practices as well – treat PLM as a mission-critical system.

User Training and Change Management: Don't skimp on training. All user groups should have training tailored to how they will use PLM (design engineers get CAD integration training, procurement team gets how to extract BOM and supplier info, etc.). Provide quick reference guides and in-system help. Possibly have power users or champions in each department who can assist others. Collect user feedback after launch and iterate on the configuration if needed to improve usability. Communicate success stories (e.g., "since PLM, our ECO cycle time went from 10 days to 3 days, saving X cost") to reinforce adoption. Culturally, encourage using the PLM by perhaps even turning off old ways (for example, disable old network drive access once files are in PLM to force the switch in a controlled manner).

Continuous Improvement: Implementing PLM is not a one-time project, but establishes a platform that should evolve. After initial go-live, measure key performance indicators (KPIs) such as the number of changes processed, time per change, reuse rate, etc. Use these to identify further improvement opportunities (maybe the change workflow is still too slow – analyse and adjust it, or maybe more automation can be introduced). As new PLM features or updates come, evaluate and adopt those that add value (for example, a new AI-based module becomes available – consider if it can help your processes). Keep the PLM aligned with business changes – if the company shifts to a new business model (say, increasing focus on services), extend PLM to manage service knowledge or assets.

To illustrate, companies that have done PLM implementations often form a centre of excellence (COE) for PLM that maintains best practices and drives these continuous improvements. They ensure the system grows with the company and stays aligned with standards and goals.

In conclusion, adhering to relevant standards like STEP and ISO 9001 in your PLM strategy ensures interoperability and quality compliance, while following implementation best practices greatly increases the likelihood of a successful PLM adoption. The combination of the right technology, processes, and people approach (change management) will allow a PLM system to deliver its potential in digital manufacturing truly. PLM, implemented well, becomes a competitive advantage, enabling faster innovation, better products, and more efficient operations in an increasingly complex manufacturing landscape. By treating PLM as a strategic initiative and not just an IT deployment, companies can support a sustainable digital thread that carries them through the current Industry 4.0 revolution and beyond, with agility and compliance built in by design.

Machine Learning and Artificial Intelligence in Digital Manufacturing

Digital manufacturing is being revolutionised by the infusion of Artificial Intelligence (AI) and Machine Learning (ML) technologies. Manufacturers increasingly view AI/ML as strategic tools for efficiency and innovation; in fact, 85% of companies have already invested or plan to invest in AI/ML in the near term[132]. AI-driven solutions are being applied from the factory floor to the supply chain, enabling data-driven decision-making and automation at unprecedented levels. The impact is evident in early adopters: AI is rapidly transforming the factory floor, accelerating the shift toward smarter, more efficient operations. From predictive maintenance to quality control, AI-powered systems are optimising production lines, driving cost savings and reducing emissions[133]. In essence, AI and ML are becoming foundational to modern manufacturing competitiveness.

AI vs. ML – Definitions in Manufacturing:

Artificial Intelligence is a broad field encompassing techniques that allow machines to mimic human-like intelligence, from problem-solving and perception to decision-making. Machine Learning is a subset of AI focused specifically on algorithms that learn from data. In a manufacturing context, AI can refer to any intelligent system (including expert systems, robotics, or rule-based automation). In contrast, ML typically means data-driven modelling – using historical sensor readings, production data, etc., to learn patterns and make predictions without explicit programming [134, 135]. In practical terms, AI is the umbrella concept ("smart machines"), and ML is one of the primary means to achieve that smartness by training models on data. For example, an AI system on a production line might include a computer vision module to inspect products and a knowledge-based system for scheduling; the vision module likely uses ML (trained on images of defects vs. everyday products) to classify items. The key difference lies in approach: AI may use a variety of techniques (including hard-coded rules or optimisation algorithms) to emulate intelligent behaviour, while ML relies on learning from examples. In manufacturing, this means ML algorithms ingest large volumes of production data to, say, predict equipment failures or product quality. In contrast, a broader AI solution might also encompass robotics or control logic that acts on those predictions. In short, AI refers to a large umbrella of technologies that include machine learning, machine vision, and deep learning [135] (among others), and ML is the workhorse under that umbrella that finds patterns in data and enables many of the "smart" capabilities on the factory floor.

Key AI/ML Subfields Relevant to Manufacturing

AI and ML comprise a spectrum of sub-disciplines. In digital manufacturing, several subfields of AI/ML play critical roles:

Supervised Learning: Supervised ML uses labelled historical data to train predictive models[136]. This is widely used in manufacturing for tasks like predictive maintenance (training on past machine sensor readings labelled with whether a failure occurred) and quality classification (training on images of products labelled as "pass" or "defect"). By learning from examples of known outcomes, supervised models can predict future outcomes, e.g. predicting the remaining useful life of a machine or classifying a weld as good or bad. Standard techniques include regression (for numerical predictions, like forecasting demand or tool wear) and classification (for categorical predictions, like defect detection)[136]. Most AI applications in factories today are based on supervised learning, since many use cases (quality inspection, failure prediction, yield optimisation) have historical examples to learn from.

Unsupervised Learning: In contrast, unsupervised ML finds hidden patterns or groupings in data without explicit labels[136]. Manufacturing data from production processes can be vast and unlabeled – unsupervised techniques like clustering and anomaly detection help make sense of it. For example, an unsupervised algorithm might analyse sensor data from hundreds of machines to cluster similar operating behaviour, identifying anomalies that could indicate a novel failure mode or a drift in process behaviour. This is useful for anomaly detection in equipment (flagging when a machine's sensor readings look very different from normal) or for grouping products/customers for demand segmentation. Unsupervised learning can uncover patterns that engineers weren't explicitly looking for, providing insights into process variation or emerging quality issues. It's essentially a tool for exploratory data analysis in manufacturing operations, revealing structure in unlabeled IoT and production datasets.

Reinforcement Learning (RL): RL is a learning paradigm where an agent learns to make decisions by trial and error, receiving rewards for desirable outcomes. In manufacturing, RL is gaining traction for process control and optimisation. An RL agent can adjust process parameters (temperature, speed, etc.) in a simulator or controlled setting, learning policies that maximise yield or throughput. For instance, researchers have applied RL to complex control problems like steel rolling and chemical process control, where the system learns to optimise quality metrics over time. One real-world use is dynamic production scheduling – an RL system can learn the best scheduling or routing of jobs in a factory to minimise delays or energy usage. While still emerging, RL has shown it can optimise manufacturing processes, leading to increased production efficiency and reduced waste[137]. A notable example is using deep RL to optimise multi-robot coordination or adaptive machine calibration in real time, where conventional heuristics fall short. As computing power and simulation capabilities grow, RL is poised to tackle more autonomous control tasks on the shop floor.

Computer Vision: Computer vision (often powered by deep learning, a subset of ML) enables machines to interpret visual data (images or video). This is highly relevant in manufacturing for automated visual inspection and quality control. AI vision systems use cameras on production lines to examine products for defects, measure dimensions, or check assembly steps – far faster and more consistently than the human eye. Modern vision models can be trained to detect tiny cosmetic defects, alignment issues, or foreign objects. The benefits include higher consistency and the ability to catch defects that humans might miss due to fatigue or subtlety. For example, a manufacturer implemented a vision system and achieved an 80% reduction in quality inspection lead time with more than 85% prediction accuracy[138]. Vision algorithms also guide robots (e.g. for bin-picking, locating parts), ensure safety (detecting if a person is too close to a robot or not wearing protective equipment), and read text via OCR (for automatically reading labels or instrument gauges). In short, computer vision gives "eyes" to factory machines, critical for high-speed inspection and flexible automation.

Natural Language Processing (NLP): Manufacturing operations generate a wealth of unstructured text data – maintenance logs, operator notes, quality reports, equipment manuals, sensor alerts, etc. NLP techniques allow AI to parse and derive insights from text. One valuable application is analysing maintenance work order logs and technician notes: NLP can identify recurring failure descriptions or link textual symptoms to machine faults. NLP transforms maintenance operations by analysing textual maintenance logs, technician notes, and equipment documentation to identify patterns and predict failures before they occur [139]. For example, suppose multiple technicians report "unusual vibration and slight burning smell" in motor maintenance notes. In that case, NLP can cluster those and correlate with a specific impending failure that might not be evident from sensor data alone[139]. This supplements sensor-based predictive maintenance with human-recorded context. NLP is also used for documentation search and retrieval (enabling workers to query manuals or past solutions in natural language) and for chatbots or voice-assistants on the factory floor (answering questions like "Why is machine seven down?" by pulling relevant data). Another emerging use is in supply chain and ERP data – NLP can read procurement contracts, emails or supplier reports to flag delays or risks. Overall, NLP helps unlock "tribal knowledge" and text-heavy data that was previously hard to integrate into digital systems, making manufacturing more information-driven.

Generative AI: Generative AI refers to models (like Generative Adversarial Networks or large language models) that can create new content, whether designs, images, or text, resembling the data they were trained on. In manufacturing, generative AI opens new frontiers in design and process innovation. One use case is generative design, where AI algorithms generate optimal product designs based on specified constraints (weight, strength, cost, etc.). For example, General Motors used generative design to reinvent a seat-belt bracket, and the AI-designed bracket ended up 40% lighter yet 20% stronger than the original, consolidating what was previously an 8-part assembly into a single 3D-printed component[140]. This highlights how generative approaches can produce highly efficient, non-intuitive designs that human engineers might not conceive. Generative AI can also produce synthetic data for simulations or train vision systems (e.g. generating images of defective products to augment limited real data). In production, generative AI aids in creating "digital twins" – virtual replicas of equipment or processes – by learning the behaviour of a system and then generating simulated scenarios (for example, simulating how a process would behave under different settings to find optimal conditions). On the textual side, large language models (LLMs) can automatically generate work instructions, summarise technical documents, or even serve as conversational assistants to engineers (answering questions using a plant's knowledge base). Generative AI is relatively new in manufacturing but is expected to play a growing role in design automation, rapid prototyping, and knowledge capture in the coming years.

Each of these subfields, supervised learning, unsupervised learning, RL, computer vision, NLP, and generative AI, provides different capabilities, and they are often combined in manufacturing solutions. For instance, an "AI system" for quality might use computer vision (CV) with a supervised deep learning model to detect defects, plus an NLP component to read operator comments about defects, and perhaps a reinforcement learning agent to adjust machine settings in response to the defects detected. Together, these AI/ML technologies form the toolkit that powers intelligent manufacturing.

Applications of AI/ML in Manufacturing

Building on those technologies, manufacturers are deploying AI/ML across a wide range of use cases. Some of the primary applications of AI and ML in manufacturing include:

Predictive Maintenance and Asset Health

Unplanned downtime is a notorious cost driver in manufacturing. AI-driven predictive maintenance addresses this by forecasting equipment failures before they happen, so maintenance can be performed just-in-time. ML models analyse streams of sensor data from machines – vibration, temperature, pressure, motor currents, etc. – to detect the subtle patterns that precede a failure (patterns often undetectable by humans or simple thresholds). Instead of relying on fixed maintenance schedules or static SCADA alarms, predictive maintenance uses algorithms to predict the subsequent failure of a component, allowing technicians to intervene at the optimal time[141]. For example, an ML model might learn that a combination of rising spindle vibration and fluctuating motor current is an early warning of a bearing failure on a CNC machine, triggering a replacement before catastrophic breakdown. Predictive maintenance systems often estimate the Remaining Useful Life (RUL) of components so that parts are utilised fully but not run to the point of breakage [141]. The benefits are dramatic: reduced unplanned downtime, extended equipment life, and lower maintenance costs. One global chemicals company reported that IoT-based predictive analytics (with AI models on sensor data) reduced equipment downtime by more than 50%[133]. To implement this, AI models are typically integrated with a CMMS (Computerised Maintenance Management System) or alerting system, so that when a prediction indicates risk, a maintenance work order or notification is automatically generated. This shifts maintenance from reactive to proactive. The success of predictive maintenance relies on having good historical failure data to train models (a challenge for rare failures). Still, even unsupervised anomaly detection can flag unusual equipment behaviour for inspection. Overall, this application of AI is one of the most widespread in manufacturing due to its clear ROI in reducing costly downtime.

Quality Inspection and Defect Detection

Ensuring product quality is paramount in manufacturing, and AI has become a game-changer in automated quality inspection. Traditional quality control involved either manual inspection (which can be slow and inconsistent) or basic sensor checks. Now, AI-powered computer vision systems can inspect products in real-time with high accuracy. Cameras combined with deep learning models can detect surface defects, dimensional errors, misassemblies, discolourations, you name it – at high speed. These systems "learn" from labelled images of good vs. bad products, and then can spot defects that are difficult to catch with the naked eye. For instance, an AI vision system on a packaging line can instantly reject items with even minor label misprints or cosmetic damage that humans might overlook after hours of repetitive checking. The consistency of AI inspection improves quality and frees up human inspectors for more complex tasks. As an example, a German manufacturing conglomerate deployed an AI vision solution and achieved an 80% reduction in inspection time while maintaining over 85% accuracy[138]. AI doesn't get fatigued, so it can inspect 100% of products (instead of just samples) and can do so at production speed (using high-speed cameras and parallel processing). Beyond final product inspection, computer vision is also used in in-process quality control – for example, checking if components are correctly placed on a circuit board before soldering, or if a weld bead's shape is within tolerance. AI can also integrate measurements from sensors and images to predict quality issues (so-called predictive quality). By analysing process data (temperatures, pressures) alongside vision data, ML models might predict a quality deviation before the product is finished, allowing adjustments on the fly. The result is lower defect rates, less scrap and rework, and higher customer satisfaction. Leading manufacturers have reported significant drops in defect rates by combining AI visual inspection with process analytics – one company noted a 53% reduction in poor quality rate after deploying AI across various quality and equipment monitoring applications[133].

Process Optimisation and Adaptive Control

Manufacturing processes involve countless parameters that must be tuned for optimal performance – temperatures, speeds, feed rates, timings, chemical concentrations, etc. Traditionally, these are set by human engineers based on experience or one-factor-at-a-time experiments. AI/ML offers a way to optimise processes in a multidimensional, dynamic way. By mining historical production data, ML models can identify what parameter settings lead to the best outcomes (highest yield, lowest cycle time, least energy use). More dynamically, reinforcement learning or advanced optimisation algorithms can be employed to adjust process settings in real-time continuously. One Lighthouse manufacturing site described deploying a machine learning-powered control system that adjusts parameters in real time, reducing scrap and preventing defects, which led to a 12.5% material cost savings in a sheet metal forming process [133]. Another example comes from injection moulding: a closed-loop AI system used a neural network to optimise the valve gate control, analysing thousands of data points and improving cycle time by 18%[133]. These examples highlight how ML can handle complex interactions between variables – something very hard to do with manual tuning. Additionally, AI can help in multi-objective optimisation – finding process settings that achieve a balance (e.g., maximising throughput without sacrificing quality beyond a limit). In continuous process industries (like chemicals or pharmaceuticals), AI-driven process digital twins are used to simulate and optimise conditions for yield and quality[133]. Engineers can ask "what-if" questions to the AI: e.g., what if we increase the temperature by 5°C, will the defect rate drop? The AI twin can predict outcomes, guiding optimal setpoints. In effect, AI adds a layer of adaptive control on top of traditional PID loops or control systems, monitoring sensor data and tweaking setpoints within safe limits to keep the process at peak performance. These adaptive systems make production more resilient and self-correcting, reducing waste (since the process stays in spec) and improving output. Over time, such AI systems can even handle changes like a new raw material batch or environmental conditions by recognising patterns and compensating (where a fixed control recipe might falter). Process optimisation through AI thus leads to higher yields, lower scrap, and faster cycle times, directly impacting the bottom line.

Supply Chain Forecasting and Demand Planning

AI's influence isn't limited to the factory floor – it extends into supply chain management and planning, which are critical to manufacturing operations. Machine learning forecasting models can analyse historical sales, market indicators, and even external variables (like economic data or weather) to produce more accurate demand forecasts. These forecasts help manufacturers with production planning: how much of each product to make and when, to meet customer demand without overstocking or understocking. Traditional forecasting might use simple statistical methods, but ML can capture complex patterns (seasonality, trends, sudden changes) better. For example, AI demand planning at a consumer goods manufacturer might incorporate social media sentiment or online search trends as features to predict a spike in demand for a product. On the supply side, AI can improve supply chain visibility and resilience. ML models can predict supplier delivery delays (by analysing supplier data, logistics trends, and even news) and optimise inventory levels accordingly. By crunching large datasets, AI finds the sweet spot that minimises inventory carrying cost while avoiding stockouts. One practical application is dynamic safety stock optimisation: an ML model continuously learns the variability in supply and demand and adjusts safety stock buffers in real time, rather than a static policy. Another application is production scheduling (which ties into supply chain and the factory): AI algorithms (including ML and heuristic optimisation) generate efficient production schedules that account for machine availability, raw material arrivals, and due dates. This often involves solving complex combinatorial problems – something AI can do faster by learning from prior schedules or using intelligent search (even RL has been explored for adaptive scheduling). The benefit is a schedule that better meets delivery targets and maximises asset utilisation. One case study noted that AI-based scheduling and dispatch in a dairy production supply chain increased inventory turnover by 73% and boosted operational efficiency by 8%[133]. Moreover, AI-driven demand forecasts feed into procurement and capacity planning decisions, ensuring the entire supply chain is more synchronised. All of this leads to lower lead times, reduced inventory costs, and the ability to respond faster to market changes – a crucial advantage in today's volatile supply chains.

Energy Efficiency and Sustainability

Manufacturing is energy-intensive, and improving energy efficiency is a key operational and environmental goal. AI is increasingly used to monitor and minimise energy consumption in production. Machine learning models can analyse energy usage patterns of equipment and find opportunities to reduce consumption during idle times or optimise process energy. For instance, an AI system might learn that specific machines can go into a standby mode for a few minutes between production cycles without affecting output, thus saving power. Or in process industries, AI might optimise heating/cooling cycles to use just enough energy for the required output quality. At a chemical manufacturing firm, AI-driven analytics managed energy consumption and achieved a 20% reduction in Scope 1 emissions by optimising processes for energy efficiency [133]. AI can also help in demand response – adjusting factory energy usage in response to grid signals or peak pricing (saving costs and stabilising the grid). Beyond direct energy use, AI contributes to sustainability through waste reduction (by improving yield and quality, as noted) and emissions monitoring. AI models can analyse emission sensor data to detect anomalies or inefficiencies that cause excess emissions, enabling quick corrective actions. There's also increasing use of AI in predictive maintenance for energy equipment (like HVAC systems in a plant or compressed air systems) to ensure they run efficiently. Some factories employ AI to decide when to run specific energy-heavy processes, scheduling them at times of lower grid carbon intensity or cost. In sum, by making production more intelligent and responsive, AI helps cut energy waste – a win-win for cost reduction and environmental impact. Many advanced facilities now include energy KPIs in their AI optimisation targets, effectively creating a "green AI" approach to manufacturing operations.

Autonomous Systems and Robotics

Manufacturing has a long history of automation and robotics, but AI is pushing this into a new realm of autonomy and human-machine collaboration. Traditional robots are typically pre-programmed to perform specific tasks in structured environments. With AI/ML, robots and autonomous vehicles in factories are becoming more flexible and intelligent. For example, autonomous mobile robots or AGVs (automated guided vehicles) in a warehouse can use AI-based vision and path planning to navigate dynamic environments, avoid obstacles, and collaboratively move materials, effectively acting as self-driving vehicles on the shop floor. In assembly operations, collaborative robots (cobots) equipped with AI vision and force sensors can work alongside humans safely, adjusting their actions in real-time. These cobots can take over repetitive or ergonomically challenging tasks (like continuous picking, fastening, or inspection), while humans handle tasks that require dexterity or decision-making. The new generation of AI-powered collaborative robots can work safely alongside humans, taking on repetitive tasks while employees focus on more complex and creative work[142]. This collaboration boosts productivity and reduces injury risk for workers. AI also enhances robotic precision and adaptability – for instance, a robot arm with an AI vision system can identify random parts on a conveyor and pick them up (a task previously arduous unless parts were exactly fixtured). We are also seeing AI use in autonomous guided forklifts, drones for inventory counts, and robotic welding/fabrication systems that adjust parameters on the fly for perfect welds. In process industries, autonomous control agents (using RL or advanced analytics) can run an entire production line with minimal human intervention, essentially acting like an autonomous operator that observes, decides, and acts. The ultimate expression of this is the "lights-out factory" concept – a production facility that runs with nearly zero direct human labour, relying on AI, robots, and automation. While fully lights-out operation is still rare (confined to certain high-volume, stable processes), we're moving closer to that with AI handling more decisions. Even when humans remain in the loop, their role shifts to supervisors of fleets of AI-driven machines. As an example, Siemens reports that in one of its electronics factories, AI-enabled robots handle pick-and-place tasks in fully automated lines, reducing automation costs by 90%, while manual workers are empowered with AI-guided systems that help them make decisions, thus enhancing overall productivity and quality[133]. AI is not about replacing all humans – it's about levelling up automation to handle complexity and variability, with humans and AI working in tandem. The result: faster operations, higher throughput, and a safer work environment where machines do the dull and dangerous and humans do the strategic and creative.

Human-Machine Collaboration and Decision Support

Rather than operate autonomously in isolation, the most effective AI deployments in manufacturing often involve humans and AI working together. AI can serve as a decision support tool for human operators, engineers, and managers. For example, an AI system might analyse in-process data and suggest to a line operator: "There is a 95% chance that Machine X will fail within the next hour due to high spindle vibration – consider switching it to a backup." The human can then take that recommendation and act, armed with foresight that wasn't available before. This kind of augmented decision-making extends from the shop floor to the executive level – AI dashboards can highlight production bottlenecks or quality hotspots in real-time, so managers can allocate resources or adjust plans quickly. In maintenance, technicians can use AI-driven diagnostic tools that, given some sensor readings and symptoms, suggest likely root causes (almost like an expert advisor built from historical data). Another area of collaboration is augmented reality (AR) combined with AI: A maintenance worker wearing AR glasses could see AI-generated annotations on a machine (e.g., "Valve #3 temperature abnormal" or step-by-step instructions retrieved by an AI from the manual) as they work. AI can also facilitate training – new operators can learn complex procedures with AI tutors that guide them and answer questions. Importantly, companies have found that involving the workforce with AI is critical: AI functions best with humans as the core decision-makers within AI-enhanced processes[135]. Rather than black-box automation, successful manufacturers keep workers in the loop, upskilling them to work with AI tools. This not only improves trust and adoption but often yields the best outcomes, combining human expertise with AI insights. A practical example is quality control: if an AI vision system flags borderline cases, those can be routed to a human inspector for final judgment. So the AI handles routine cases and the human handles exceptions, resulting in higher overall accuracy than either alone. This hybrid model is becoming the norm. AI becomes like a "co-pilot" to the human worker, much as advanced driver-assist systems in cars help drivers, AI assists factory personnel. The benefit is faster and more informed decisions and actions. Additionally, human workers are relieved of tedious data crunching and monitoring and can focus on improvement and innovation. Many organisations are explicitly focusing on this synergy (sometimes referred to as Industry 5.0, where the emphasis is on human-centric automation). By designing AI systems that augment rather than replace human capabilities, manufacturers achieve both productivity gains and workforce acceptance – truly the best of both worlds.

AI-powered robotic arms are performing packaging tasks on a production line. Advanced computer vision and control algorithms enable such robots to adapt to different products and work safely and precisely, illustrating how AI-driven automation is enhancing manufacturing processes.

Integration of AI/ML with Manufacturing Systems

To realise these applications, AI/ML solutions must be tightly integrated with existing manufacturing IT and OT systems. A factory is a complex environment with multiple software and hardware systems – Manufacturing Execution Systems (MES), Enterprise Resource Planning (ERP) systems, Product Lifecycle Management (PLM) tools, SCADA (Supervisory Control and Data Acquisition) systems, and increasingly, Industrial IoT platforms. Successful AI deployments act not as standalone black boxes but as connected parts of this ecosystem:

MES Integration: The MES is the system that orchestrates shop-floor activities (production schedules, work orders, tracking work-in-progress). Integrating AI with MES allows real-time optimisation and execution of AI insights. For example, if an ML model predicts a machine failure, the AI can signal the MES to adjust the production schedule (rerouting tasks or scheduling maintenance). Conversely, MES provides AI with valuable data – production counts, operator inputs, and downtime reasons – that can be used in training models. The integration of AI and MES is "revolutionising the industry" by creating adaptive production systems, where the plan can change on the fly based on AI predictions[143]. Some modern MES/MOM (Manufacturing Operations Management) suites even embed AI analytics dashboards and decision engines directly. Seamless integration with ERP and supply chain systems is also facilitated via MES as the bridge[144]Meaning AI can help synchronise factory-floor decisions with business-level considerations (orders, inventory). For instance, an AI might detect a quality issue and, through MES and ERP, automatically put a hold on shipping specific batches and initiate replenishment orders for new material. Tight MES-AI coupling leads to a more flexible, intelligent, and adaptive plant, essentially creating the execution layer of the "smart factory."

ERP and Supply Chain Systems: ERP software handles company-wide functions like inventory, procurement, and order management. Integrating AI here means connecting ML models for demand forecasting or supply optimisation into the ERP workflows. An example integration is using an AI forecast to automatically adjust production planning parameters in the ERP's MRP (Material Requirements Planning) run. Or an AI-driven supplier risk score might be integrated into the ERP's purchasing module, flagging certain suppliers and suggesting alternatives. The key is that ERP has the data (historical sales, inventories, suppliers, costs), and AI provides the advanced analytics. Many manufacturers choose to implement AI in parallel (like a cloud AI service) but feed it data from ERP and then loop back the outputs. Others are using ERP-embedded AI modules now offered by major ERP vendors. Integration challenges include data consistency and real-time data flow, which is why some companies build data lakes or warehouses that extract and consolidate data from ERP, MES, etc., to serve as a single source for AI analysis. Nonetheless, a well-integrated AI can enable closed-loop supply chain management – e.g., a drop in forecast demand (predicted by AI) propagates to production reduction and raw material order adjustments automatically through ERP and procurement systems.

PLM and Engineering Systems: PLM systems manage the product lifecycle from design to engineering change to end-of-life. AI integration here is about feeding back manufacturing data to design and vice versa. For instance, quality data and field performance data can be aggregated (using AI analytics) and pushed into PLM to inform design improvements – a practice known as Closed-Loop PLM. If AI finds that a particular component often causes failures in the field, the design team can be alerted via PLM workflow to redesign that part. During design stages, AI-driven generative design tools (discussed earlier) might be integrated with PLM/CAD software, so that engineers can seamlessly generate AI proposals and evaluate them. AI is transforming each phase of PLM – from conception (where generative design suggests thousands of design variants) to design (where AI-powered simulation tests designs virtually with digital twins) to production (where PLM connects with MES for manufacturing feedback)[145]. As an example, digital twin simulations in the design phase can ensure a product is manufacturable by simulating the production process in a virtual replica of the factory (taking data from MES/SCADA)[145]. Thus, integration between AI tools and PLM ensures that products are designed with manufacturing insights and that production benefits from AI without violating design intent. As products become smarter (with embedded sensors, etc.), PLM may also incorporate AI to manage product data and configurations over time (like analysing IoT data from shipped products to schedule proactive service, an overlap of PLM and after-sales service enabled by AI).

SCADA and Control Systems: SCADA systems and PLCs are at the heart of real-time control in plants, providing the interface to sensors and actuators. Traditionally, SCADA operates on predefined logic and threshold-based alarms. By integrating ML algorithms into this layer, manufacturers achieve far more nuanced control and monitoring. For example, rather than a SCADA alarm that simply trips when temperature > X, an ML model might consider the combination of temperature and pressure trends to detect anomalies. Instead of static human-coded thresholds and alerts, AI allows dynamic analysis of the complex behavioural patterns of machinery[141]. One approach is to run ML models at the edge (on industrial PCs or controllers) that subscribe to the SCADA data streams. If the ML detects an anomaly or optimal adjustment, it can send a command to the PLC or trigger an alarm in the SCADA HMI with additional diagnostic info. This IT/OT convergence (information tech + operational tech) is a big theme: many plants are connecting PLC data to cloud ML services (through IIoT gateways) and then feeding results back down to control systems. For integration to succeed, latencies must be low (especially for control) and reliability high, which is why sometimes simpler ML models are deployed on the edge devices near the SCADA, with heavier analytics in the cloud for non-real-time insights. Modern control platforms are starting to include AI capabilities natively. For instance, some DCS (Distributed Control System) vendors offer AI modules that plug into the control loops. The outcome is smarter control processes that maintain optimal settings and quickly detect when something is starting to go wrong, beyond what any single sensor threshold could catch.

Industrial IoT (IIoT) Platforms: IIoT platforms (like PTC ThingWorx, Siemens MindSphere, GE Predix, etc.) are used to connect and manage the myriad devices in a factory, collect data, and often provide analytics frameworks. They are natural allies of AI/ML, since they handle the data ingestion and device management that AI systems need. Integrating AI and machine learning with IIoT platforms enhances data analysis and predictive capabilities[146]. Typically, sensor and machine data flow into the IIoT platform; from there, it can be fed into ML models to derive predictions, which are then sent to users or machines. Many IIoT platforms have built-in modules for anomaly detection, predictive maintenance, etc., leveraging ML. A concrete integration example: an IIoT platform collects vibration data from all motors in a plant and uses a cloud-based ML model to predict failures; if a prediction crosses a threshold, the platform triggers an event that notifies maintenance and also updates a dashboard in the control room. Another example is using an IIoT platform to deploy ML models to the edge. Manufacturers often require some computations to happen on-site (for latency and privacy), and IIoT platforms can push an AI container or model update to an edge gateway next to a machine. The data lake approach is also common: streaming all factory data into a central data lake (on cloud or on-premise cluster) where data scientists can train AI models, combining datasets (SCADA historian data, MES records, quality measurements, etc.). An "Operational Data Lake" was created by one company to integrate data across 50 plants, enabling delivery of AI/ML interventions to improve yield and throughput[133]. This highlights that data integration is foundational – without unified, accessible data, AI cannot deliver value. Therefore, investing in IIoT platforms or data infrastructure is a prerequisite to AI at scale. The integrated outcome is a connected factory nervous system: raw data comes in from machines via IIoT, AI brain processes it (sometimes at the edge, sometimes centrally), and the insights flow out to MES/ERP/operators, closing the loop.

In summary, integrating AI/ML into manufacturing systems means embedding intelligence into the digital fabric of the factory. Rather than bolting on an AI in isolation, the trend is to have AI woven through MES, ERP, PLM, SCADA, and IIoT layers. This integration allows AI insights to be actioned (not just observed) – whether automatically adjusting a process, triggering a maintenance order, or informing an operator. Done correctly, it yields a unified system where data flows freely and decisions are made at the right level (cloud vs edge, human vs machine). Of course, integration is often technically challenging due to proprietary protocols and legacy equipment, but progress in standards (e.g., OPC UA for industrial communication[147]) and modern APIs is easing the interoperability issues. Many manufacturers address integration incrementally: start with one line or one use-case, connect the necessary systems, prove value, then scale out connectivity plant-wide. As the backbone solidifies, AI can be layered on more broadly.

Data Requirements and Infrastructure for Industrial AI

Data is the lifeblood of AI/ML, and manufacturing environments generate enormous amounts of it, from IIoT sensor streams to production logs. However, harnessing this data for AI requires careful attention to data requirements and infrastructure:

Data Quantity and Quality: Effective ML models often need large volumes of historical data. For example, a predictive maintenance model improves with years' worth of sensor readings and dozens of examples of failures. In manufacturing, however, data can be scarce or of poor quality for specific events (e.g., a catastrophic failure might have only happened once). Moreover, plant data may be noisy or error-prone – sensors malfunction, or data might be manually entered with mistakes. A study of ML in manufacturing noted challenges like lack of data quality, heterogeneity, and data imbalance as major hurdles[148]. Ensuring data quality (clean, accurate, consistent data) is thus critical. This can involve spending significant time on data cleaning, outlier removal, and aligning data from different sources (e.g., making sure the timestamp of a sensor reading matches the correct production batch from MES). Many practitioners observe that data preparation is 80% of the effort in industrial AI projects. Dealing with labelling is another issue: supervised learning needs labelled examples (like failure/no-failure labels, or defect types on images). Labelling manufacturing data can be tedious – it might require domain experts to label thousands of pictures of defects, or to mark exactly when a machine failed in sensor logs. Some firms employ data annotation services or develop internal tools to streamline labelling. Techniques like active learning (where the model helps identify which data points to label) and transfer learning (using models pre-trained on similar data) are often used to mitigate limited labelled data [149]. Additionally, data bias must be watched – if your historical data mostly comes from operating one product type, the model may not generalise to a new product. A best practice is to continuously update and retrain models as new data comes in (ensuring distribution shifts are captured). In short, having "clean, meaningful, high-quality data is critical for the success of AI initiatives"[147], so companies must invest in data engineering up front.

Data Pipelines and Real-Time Streaming: Manufacturing AI often needs data in real or near-real time (for instance, to predict a failure a few seconds before it happens, or to adjust a process promptly). This demands robust data pipelines that can stream data from the shop floor to the AI models with low latency. These pipelines might use protocols like MQTT or OPC UA to gather sensor data, and then push it through an edge gateway or up to the cloud. For less time-sensitive analyses (like daily production reports or weekly yield predictions), batch data pipelines (periodically ETLing data into a data lake) suffice. Many organisations create a layered architecture: raw data from PLC/SCADA goes into a historian database (common in process industries), from which it's then forwarded to cloud storage for ML training; simultaneously, a subset of critical signals is streamed live to an edge ML system for instant analysis. The pipeline architecture must also handle data from different sources – machines, quality inspection devices, environmental sensors, MES/ERP databases – merging them via standard timestamps or IDs. This heterogeneity is non-trivial: for example, you might need to join a temperature time-series from a furnace PLC with the batch ID from MES and the final lab test result from a LIMS (Laboratory Information Management System) to train a model relating process conditions to product quality. Data pipelines thus involve significant integration and context-building. Technologies like Apache Kafka (for streaming), REST APIs, and IIoT middleware are commonly employed. The goal is to achieve a single, unified data layer feeding the AI. As one company did, connecting all plants to an Operational Data Lake to get a real-time, integrated view of data and deliver AI interventions (133), others aim to centralise data for consistency. Real-time needs also introduce the concept of data latency. If an AI is to act in real time, the end-to-end latency from data generation to decision must be within the required window (e.g., a few milliseconds for high-speed control, or a few minutes for maintenance decisions). This sometimes leads to doing AI at the "edge" (on-premise) for speed, which we discuss next.

Edge Computing vs. Cloud Computing: Manufacturing data can be processed in the cloud (off-site servers/data centres) or at the edge (on-site computers, on the device itself, or nearby). Edge computing is often crucial for manufacturing AI because of latency, bandwidth, and reliability concerns. Edge computing processes data locally at the edge of the network – near the machines and sensors where data is generated – which accelerates decision-making and reduces the need to send vast data to a distant centre (150). For example, a vision inspection system might generate 100 images per second; sending all to the cloud would be impractical, so an edge GPU device runs the vision AI model on-site and only sends aggregated results to the cloud. Edge computing also keeps operations running even if the internet connection drops – a critical safety and uptime consideration. Furthermore, it can address data privacy/IP concerns (sensitive production data might not leave the factory). On the other hand, cloud computing offers virtually unlimited compute power and storage, useful for training large models or aggregating data across multiple sites. Many manufacturers adopt a hybrid architecture: initial model development and heavy analytics happen in the cloud (where you can crunch years of data, use significant clusters, etc.), but the resulting model is then deployed to the edge for execution on the line. For instance, you train a predictive maintenance model on cloud servers using data from 10 factories, but then you push that model to each factory's local server to run in real time on incoming sensor feeds. Modern IIoT platforms support this model deployment. Edge devices might include industrial PCs, embedded controllers, or even AI chips on the equipment. Balancing edge and cloud is a design decision – the rule of thumb: keep what needs quick action and high bandwidth at the edge, and use the cloud for global optimisation, fleet learning, and heavy computation. With the rise of 5G private networks in factories, the line between edge and cloud blurs a bit (since 5G can transmit extensive data quickly on-site), but the fundamental trade-offs remain. The end infrastructure often looks like: machines -> local network -> edge gateway (for quick ML tasks) -> cloud data lake (for storage and big analytics) -> cloud ML services (training/improvement) -> back to edge (updated models). This ensures low-latency control (thanks to the edge) and continuous learning and improvement (thanks to the cloud's power).

Data Lakes and Storage: Given the variety and volume of manufacturing data, companies build data lakes or warehouses as central repositories. A data lake can store raw data in its native format (e.g., time-series, images, text logs) and allows flexible retrieval for different analytical needs. Manufacturing data is often time-series heavy – specialised time-series databases or historians (like OSIsoft PI) are used. Still, increasingly those feeds are duplicated to cloud storage (S3, Hadoop, etc.) to enable AI processing with big data tools. A well-structured data lake will tag data with contextual metadata (like which plant, line, product, batch it belongs to) so that data scientists can query, say, "give me all temperature profiles and final quality outcomes for product X in the last 2 years" to train a model. Data lakes also allow mixing data – merging, for example, MES transactional data (which might be stored in SQL databases) with unstructured data (like maintenance logs or PDFs of test certificates) and IoT data. This comprehensive view is compelling for AI, as complex phenomena in manufacturing often span multiple systems (a defect could be due to a subtle combination of machine behaviour, raw material property, and operator action, requiring data from equipment, materials, and HR training logs, perhaps!). Without a unified storage, these connections would be missed. However, simply dumping data is not enough; companies need data governance to ensure data is labelled, accessible, and secure. Many are establishing data pipelines with ETL processes that clean and structure data as it lands in the lake (e.g., converting units, filtering outliers, aligning timestamps). There's also an increasing use of cloud-based AI/analytics platforms (from AWS, Azure, Google, etc.) where the data lake and ML tools coexist, simplifying the pipeline. An emerging concept is the manufacturing data mesh – treating data as a product, owned by domain teams (like a data product for "machine telemetry" or "quality results"), which are made available to AI consumers via standardised interfaces. Whether lake or mesh, the goal is to break down silos and make high-quality data readily available for AI modelling.

Edge AI Hardware: On the edge, we see specialised hardware being deployed, such as NVIDIA Jetson devices for on-site image processing, or FPGA/ASIC accelerators in vision cameras that can run neural networks directly on the camera ("smart cameras"). PLC manufacturers are also integrating AI, for example, PLCs with built-in neural network modules that can run a small ML model alongside ladder logic. This hardware evolution is part of the infrastructure that enables AI in real-time control contexts. Factories may also leverage on-premise GPU servers to handle multiple AI tasks locally (like an on-site mini "AI cloud" that might run vision for multiple lines, plus some scheduling optimisation, etc.).

In summary, data readiness is as essential as algorithm development in manufacturing AI projects. Companies must ensure they have robust data pipelines, storage, and processing capabilities – essentially an information architecture that can support AI. "Garbage in, garbage out" holds: without the correct data in the proper form, even the best AI algorithm will underperform. By investing in IoT connectivity, data lakes, and edge/cloud computing resources, manufacturers create a strong foundation where AI solutions can thrive and be scaled across the enterprise.

Benefits of AI/ML in Manufacturing

When AI and ML are effectively implemented in manufacturing, the benefits can be substantial. Key advantages include:

Increased Productivity and Efficiency: AI enables faster production cycles and better use of resources. Intelligent automation and process optimisations allow more output with the same or fewer inputs. Early adopters report boosts in productivity and efficiency – for example, Siemens achieved significant efficiency gains by optimising testing and production with ML[133]. By reducing bottlenecks and downtime and automating routine tasks, AI helps factories produce more in less time.

Reduced Downtime: Through predictive maintenance and real-time monitoring, AI minimises unplanned machine outages. This improves equipment OEE (Overall Equipment Effectiveness). Some plants have cut downtime by 50% or more by catching failures in advance[133]. Less downtime means higher uptime for production assets and less scrambling to expedite orders or idle labour, directly saving costs.

Lower Defect Rates and Improved Quality: AI-driven quality control catches defects earlier and more reliably, leading to fewer defective products escaping or needing rework. ML can also adjust processes to keep quality within spec. Midea Group, for instance, saw a 53% reduction in poor quality after widespread AI deployment[133]. Consistent quality improves customer satisfaction and reduces scrap and warranty claims.

Adaptive and Agile Processes: AI brings adaptability – processes can adjust in real time to variability (machine wear, material differences, etc.). This means manufacturing lines are more flexible and can switch products or personalise products with less manual re-tuning. It also aids in mass customisation, as AI can handle complexity that would bog down traditional automation. In essence, AI can make manufacturing processes self-optimising and resilient to disturbances (e.g., automatically compensating for a slight change in raw material properties to still produce sound output).

Informed Decision Making and Decision Support: AI provides actionable insights from data that would otherwise be overwhelming. Engineers and managers get data-driven recommendations – e.g., which process parameter most influences a quality issue, or which supplier is most reliable. This enhances human decision-making. Rather than relying purely on experience or gut feel, teams have analytics to back up decisions on everything from daily operations to strategic planning. AI-powered dashboards can synthesise thousands of data points into a clear picture (like highlighting an emerging production trend), enabling quicker and more confident decisions.

Cost Reductions: Many of the above points translate to cost savings – less scrap, less downtime, optimised inventory, reduced energy usage. Additionally, AI can reduce labour costs on mundane tasks (through automation) while often increasing labour productivity (workers focus on higher-value tasks). Some AI initiatives directly target cost, such as optimising cutting tool usage to extend tool life or reducing material waste through better process control. Cost of poor quality (COPQ) often drops with AI, as it can catch defects early or prevent them. Lean manufacturing principles get a high-tech assist from AI, eliminating wastes that were previously hidden in data.

Enhanced Safety and Risk Management: AI systems can improve safety by monitoring conditions and enforcing rules. Computer vision can detect if a worker isn't wearing proper PPE or has entered a hazardous zone, and then trigger alerts or even shut down equipment. Predictive maintenance also reduces the risk of catastrophic machine failures that could injure workers. Moreover, AI can simulate and evaluate risky scenarios (digital twins running "what-if" tests for dangerous conditions) so that issues are mitigated before actual implementation. A safer operation not only protects people but also avoids downtime and liability from accidents.

Workforce Augmentation and Skill Enhancement: Rather than replacing workers, AI often augments them. It takes over repetitive, strenuous, or computationally heavy tasks, freeing humans for more complex work. This can improve job satisfaction (workers engage in problem-solving and innovation instead of a dull routine). It also helps address the manufacturing skills gap: new or less experienced workers can be effective sooner by relying on AI guidance and expertise systems. Companies have found that AI tools, combined with training programs, can upskill workers and make their jobs more engaging[135]. Essentially, AI becomes a teammate – one that provides memory, calculation, and vigilance, complementing human creativity and judgment.

Sustainability and Energy Savings: As mentioned, AI helps reduce energy and raw material usage, which supports sustainability goals. It can also optimise supply chains for a lower carbon footprint (e.g., reducing expedited shipments, optimising loads). The WEF's Global Lighthouse Network shows that AI-enabled sites often set new standards for sustainability, reducing emissions and waste[133]. Sustainability is both an environmental and economic benefit – using fewer resources typically saves money, and it future-proofs operations against environmental regulations or resource scarcity.

Innovation and New Capabilities: AI can unlock new manufacturing capabilities that were previously not feasible. For example, real-time adaptive control might allow entirely new process techniques or ultra-precise manufacturing. Generative design can create products of superior performance that traditional methods wouldn't produce. AI can also enable better customisation at scale, opening new markets (like one-off personalised products made efficiently). Moreover, by analysing data across the product lifecycle, AI can suggest innovative improvements in products and processes – essentially becoming a catalyst for continuous improvement (kaizen) on steroids. Many companies find that once they incorporate AI, they discover new insights and opportunities that lead to product innovations or new business models (such as offering AI-driven "smart" services along with products).

All these benefits contribute to a significant competitive edge. Manufacturers that effectively leverage AI/ML tend to see higher throughput, better quality, lower costs, and faster responsiveness, which translates to higher customer satisfaction and market share. It's important to note that quantifying AI benefits often requires careful measurement and a bit of time (some benefits are immediate, like reduced defects; others accrue over months, like improved maintenance costs). Nevertheless, numerous case studies now document tangible ROI. For instance, a survey indicated that adopters are seeing AI as core to their tech strategy and delivering returns, second only to cloud technologies in ROI[132]. From the factory floor to the boardroom, the consensus is that AI/ML, when done right, drives value creation in manufacturing.

Challenges of Implementing AI/ML in Manufacturing

Despite the compelling benefits, deploying AI and ML in manufacturing comes with a set of challenges and obstacles that organisations must navigate:

Data Availability and Labelling: As emphasised, AI needs data, but many manufacturers struggle with having enough relevant, high-quality data. Some equipment may not have sensors in place to generate data, or historical data may not have been stored. In other cases, data exists but is siloed in disparate systems or formats that are not readily usable. Additionally, supervised learning needs labelled examples, which may require experts to spend time annotating data (e.g., labelling images with defect types, or tagging periods of machine failure in time-series data). Manual labelling of manufacturing data can be a daunting task, especially in complex environments with ambiguous failure modes[151]. For instance, if a product failed a stress test, identifying exactly which process data points to label as "faulty" can be non-trivial. Imbalanced data is another issue – critical failures are (hopefully) rare, so algorithms struggle to learn from a small number of examples. Overcoming this challenge often requires investment in sensor retrofitting (to gather new data), better data integration (so AI can access all existing data), and creative techniques to augment or simulate data. Some companies collaborate with industry consortia to share anonymised data to get more failure examples (when direct data sharing isn't possible, this could segue into federated learning, discussed later). In summary, getting the correct quantity and quality of data remains a foundational hurdle – many AI projects stall at the data stage if not enough clean data is available.

Data Quality and Management: Even when data is available, it may be dirty or inconsistent. Manufacturing environments introduce noise: faulty sensor readings, data entry errors, misalignments in time stamps, etc. Data might also be incomplete (e.g., missing values when sensors go offline). If AI learns from flawed data, it will produce flawed conclusions. Thus, ensuring data quality (through cleaning, filtering, and validation) is a big challenge. Moreover, manufacturing data is often highly heterogeneous – different formats, frequencies, units – making it hard to combine. A simple example: one system logs time in UTC, another in local time; if not reconciled, an AI model could be thrown off. Data governance practices in many factories are still maturing; unlike banks or online companies, manufacturers historically didn't treat data as a core asset, so legacy data may not have been well-kept. Changing this mindset is part of the challenge – establishing proper data management, metadata documentation, and oversight of data changes. Another aspect is data silo culture – different departments "own" their data and may be reluctant to share or change how they handle it. Overcoming organisational silos to create an enterprise-wide data strategy is often as tough as the technical cleaning itself.

Model Complexity and Interpretability: Many AI models, especially modern deep learning ones, are complex "black boxes." On a shop floor, if an AI model flags a potential issue or recommends a setting change, engineers will naturally ask, "Why?". Interpretability (or explainability) of AI predictions is crucial for trust. However, explaining a deep neural network's reasoning can be difficult. The lack of interpretability of AI/ML outputs makes it challenging to use for planning, as humans struggle to trust opaque models[152]. This is a barrier: manufacturing experts might resist an AI system's advice if they don't understand the basis, especially when decisions impact safety or costly processes. Furthermore, regulatory compliance or internal policies might require explanations. E.g., in pharma manufacturing, if an AI recommends a process change, you need to validate and explain that change for regulatory approval. To address this, techniques like SHAP values or rule extraction can be applied to models to give some explanation (e.g., "the model thinks the motor will fail mainly because vibration increased by X and temperature by Y"). Simpler models (like decision trees or regression) are more interpretable but might be less accurate than complex ones. This trade-off is often considered: sometimes, a slightly less accurate but interpretable model is preferred for critical decisions. Building user trust in AI means involving them in model development (so they see how variables correlate) and gradually proving the model's efficacy. In any case, overcoming the "black box" challenge is key – lack of trust in AI's capabilities is a significant barrier to broad adoption[147], so transparency and education must be part of the implementation.

Integration with Legacy Equipment and Systems: Factories often run on legacy machines and software that were never designed to connect to modern networks or output data in AI-friendly formats. Interoperability is thus a big challenge – how to extract data from an old CNC machine, or how to send an AI control signal to a legacy PLC that only accepts specific inputs. Many plants have a mix of old and new equipment, leading to a "Tower of Babel" of protocols. Upgrading all legacy equipment is usually infeasible due to cost. Instead, solutions like IoT retrofits (sensors added to old machines), protocol converters, or middleware that can talk to various systems are needed. This is as much an engineering project as an IT one. Manufacturing sites often have a wide variety of machines using different, sometimes competing, technologies, some running outdated software not compatible with other systems[147]. As noted, standards like OPC UA are helping by acting as a universal translator for industrial data[147]. But not everything supports such standards out of the box. Companies may need to custom-build interfaces or use edge gateways that interface with legacy devices (e.g., reading analogue signals or scraping data from a machine's HMI) to pull data into the AI pipeline. Similarly, feeding AI outputs back might require creative solutions – maybe a robot doesn't have an API, so you physically wire a signal that the AI triggers when a prediction is made. Ensuring real-time performance across these integrations is also tricky, as older systems might have delays or limited throughput. All this adds complexity, time, and cost to AI projects. It's often said that for every hour spent on the AI algorithm, several more are spent on integration engineering in manufacturing settings. This challenge can be mitigated by gradually modernising infrastructure and using modular, open systems when buying new equipment, to ease future AI integration.

Workforce Skills Gap: Implementing AI requires not only data scientists and ML engineers, but also domain experts who understand both manufacturing and data. There is a shortage of professionals who have the interdisciplinary skills to deploy AI in industry. Many manufacturing engineers are not trained in AI, and conversely, many data scientists are not familiar with manufacturing processes. Bridging this gap is a significant challenge. Companies report difficulty in hiring experienced AI talent willing to work in manufacturing settings[147]. Moreover, existing staff may fear or be sceptical of AI, requiring training and change management. Upskilling is needed at multiple levels: from plant technicians learning to trust and manage AI-driven systems, to engineers learning coding or data analysis, to data scientists learning the shop-floor realities. Some manufacturers have set up internal "AI academies" to train employees on data science basics[135]. Others bring in consultants or partner with universities. The cultural change is non-trivial: transitioning a traditional engineer into someone who relies on recommendations from an algorithm can be challenging. Manufacturing is expected to face a workforce shortage as older workers retire, and attracting young AI talent is difficult if they view manufacturing as unstimulating[147]. Companies need to make manufacturing high-tech and attractive, highlighting that AI in manufacturing is a cutting-edge field, to recruit and retain talent. There's also the fear aspect: workers might worry AI will replace their jobs. Addressing that concern openly (by demonstrating augmentation, not replacement, and perhaps guaranteeing roles) is part of change management. If the workforce isn't on board or capable, AI projects can fail, regardless of technical merit.

Change Management and Adoption: The introduction of AI represents a significant change in processes and decision-making. Resistance to change is common – operators might distrust a predictive system, managers might be hesitant to override their traditional KPIs or methods. Change management is therefore a key challenge: it involves communicating the purpose and benefits of AI, training people to use new tools, and often redesigning workflows. It can also involve new roles (like a data analyst on the shop floor) and shifting responsibilities (maybe an AI scheduler takes over tasks that a production planner used to do manually, so that planner's role shifts to oversight and exception handling). Ensuring that employees are engaged rather than alienated is essential. One risk is "pilot purgatory" – companies do an AI pilot in one area successfully, but then fail to scale it because broader adoption falters. Common reasons include a lack of executive championing, middle-management resistance, or insufficient alignment with business processes. One of the most significant impacts of AI in manufacturing is on the human workforce… AI must be integrated in a way that builds trust within existing risk-averse cultures[153]. Practical steps in change management include involving end-users early (having operators help design the AI interface), providing extensive training, and phasing the rollout (perhaps running the AI in "shadow mode" first to prove its suggestions while humans still make final decisions). Celebrating quick wins and showing that AI is a tool to make their jobs easier, not to judge or replace them, helps ease anxiety. A Gartner survey found that workers who fear AI will replace their jobs are 27% less likely to stay with their employer, indicating that mishandling the narrative around AI can even lead to talent retention issues. Therefore, transparent communication and a people-first approach (sometimes termed "augmented intelligence" instead of AI) are necessary to overcome this human challenge.

Cybersecurity Concerns: As factories become more connected and reliant on AI, they also become more exposed to cyber threats. Connectivity (IIoT, cloud, remote access) increases the attack surface for hackers. A successful cyberattack on a smart factory could be devastating – it might shut down production or even cause physical damage if safety systems are compromised (e.g., the infamous Stuxnet attack on PLCs). AI systems themselves could be targeted – attackers might try to feed false data to an ML model to force it into wrong decisions (so-called adversarial attacks), or use AI tools to find vulnerabilities in the manufacturing network. Manufacturing systems have become a perfect target for cybercriminals as IIoT expands – attackers can exploit the larger attack surface[154]. There's also concern that adversaries could use AI: e.g., using AI to automatically craft malware that specifically attacks industrial controllers[153]. On the flip side, AI can aid cybersecurity (by detecting anomalies in network traffic), but here we focus on the challenge it brings: securing an AI-driven operation. Companies must invest in robust industrial cybersecurity – network segmentation, encryption of sensor data, authentication for devices, and continuous monitoring. Legacy OT systems often lack security features (some PLCs have no authentication), so adding AI without addressing security can be risky. Furthermore, regulatory and compliance requirements (like for critical infrastructure) mandate specific security standards which AI implementations must adhere to. The challenge is to balance openness for data flow with security. Practical steps include involving the IT security team from the start of AI projects, performing rigorous testing (to ensure an AI that, say, connects to the cloud doesn't inadvertently open a pathway into the control network), and keeping AI models and data access under strict permission control. Data privacy might also be an aspect if, for example, vision systems record images of employees or if manufacturing data is sensitive (like defence-related designs). Ensuring AI does not violate privacy or confidentiality is part of this challenge.

Regulatory and Ethical Issues: Some industries (pharmaceutical, aerospace, food) have heavy regulatory oversight on manufacturing processes. Introducing AI may raise questions with regulators: How do you validate a machine learning model? Who is accountable if AI makes a wrong decision that affects product safety? These are challenges being navigated now. Companies need to establish validation protocols for AI (similar to how any equipment or software is validated). For instance, the FDA in pharma is open to AI but expects robust evidence of its reliability. Ethically, manufacturers must consider the impact of AI on employment, require transparency (avoid bias if AI is used in areas like hiring or performance evaluation), and ensure AI decisions align with safety and quality above all. Many leading firms create ethical AI guidelines – Johnson & Johnson, for example, developed an ethical AI framework and upskilled teams in AI engagement[135] to ensure AI use remains responsible. Ethical use also covers fairness (e.g., an AI scheduling system should not unintentionally always give the toughest shifts to certain workers, which could happen if trained on historical bias). Additionally, environmental ethics come into play: AI can be power-hungry (training big models consumes energy), so balancing the computational load with sustainability goals is considered. Another emerging ethical aspect is data ownership – if AI is trained on data from multiple suppliers or clients, who owns the model and insights? In collaborative manufacturing networks, this can become complex. Organisations must navigate these questions, often in the absence of clear external regulations, because the technology is ahead of policy. Thus, internal governance structures for AI (data committees, AI ethics boards) are helpful to address concerns proactively.

These challenges may seem daunting, but they are being actively addressed across the industry. The key is awareness and strategic planning: anticipate these hurdles and work on mitigation plans early in the AI adoption journey. For example, allocate plenty of time for data preparation, invest in employee training programs, start with AI applications that have lower risk to build trust, and involve IT/security from day one. Many failures of AI projects can be traced to neglecting one of these human or infrastructure factors rather than the AI technology itself. By tackling data, people, integration, and governance challenges holistically, manufacturers can significantly increase the chances of AI project success.

Case Studies and Real-World Examples

Real-world implementations of AI/ML in manufacturing demonstrate clear value creation. Here are a few case studies across different industry segments.

Beko (Home Appliances): Beko, a major appliance manufacturer, leveraged machine learning to optimise process parameters in real-time on their production lines. One notable success was in a sheet metal forming process: an AI-powered control system adjusts machine settings on the fly, reducing scrap and preventing defects, which led to a 12.5% material cost savings[133]. In another use case, Beko applied a decision-tree ML model to address clinching (fastening) issues caused by varying sheet thickness, cutting those defect rates by 66%[133]. They also implemented a closed-loop AI for plastic injection moulding that analysed over 150,000 data points to optimise the process, improving cycle time by 18%[133]. Beyond the technical wins, Beko invested in training, completing 3,160 hours of AI training for staff in six months to ensure the workforce could implement and scale these solutions[133]. This holistic approach – targeting specific pain points with AI and empowering employees – resulted in significant cost savings, quality improvements, and faster time-to-market (e.g., cleaning cycle design time reduced by 46% via ML algorithms[133]).

AstraZeneca (Pharmaceuticals): AstraZeneca applied AI across drug development and manufacturing, fundamentally transforming processes. In R&D, they use predictive modelling to optimise properties of drug compounds and formulations, accelerating growth. Impressively, Generative AI and machine learning have helped reduce development lead times by 50% for specific projects[133], and even cut down the use of expensive active pharmaceutical ingredients in experiments by 75%[133] – saving cost and time. In manufacturing, AstraZeneca employs AI-driven process digital twins to optimise production conditions for yield and productivity while minimising raw material use[133]. These digital twins simulate the relationship between process settings and product quality, allowing engineers to find optimal operating points virtually. Combined with continuous manufacturing techniques, these efforts reduced manufacturing lead times from weeks to hours in some cases[133]. They also utilised GenAI to accelerate documentation for regulatory filings, cutting the time to create certain documents by over 70%[133]. This case exemplifies AI enabling both speed and quality: faster development, more efficient production, and robust compliance – all critical in pharma. It also shows AI's role in sustainability, as AstraZeneca notes using AI to analyse life-cycle data and identify emissions "hotspots" to target for carbon footprint reduction[133].

Jubilant Ingrevia (Chemicals): Jubilant, a chemical manufacturer, implemented AI and ML at scale across all production stages to boost efficiency and reduce variability[133]. They heavily use digital twins of critical assets to model and manage operations in real time[133]. Specific ML models optimise various production parameters using historical and live data, ensuring consistent quality and resource efficiency[133]. By analysing processes with AI, they achieved a striking 63% reduction in process variability[133], leading to more consistent product quality. Predictive maintenance via IoT monitoring and AI has reduced downtime by >50% at Jubilant's plants[133], significantly enhancing operational efficiency. Another noteworthy achievement is in energy management: AI-driven analytics optimised energy use, yielding a 20% cut in specific emissions and lowering operational costs[133]. To facilitate these gains, Jubilant ensured all its 50 plants are connected with an Operational Data Lake for real-time integrated data – a backbone for delivering AI solutions everywhere[133]. They also established a "Digital Centre of Excellence" with AI experts and rolled out company-wide digital training, alongside incentive programs to encourage AI innovation[133]. This case shows how a traditional industry can embrace AI to achieve leaner, greener, and more reliable operations through a combination of tech and organisational initiatives.

Siemens (Electronics Manufacturing): Siemens, which is both a user and provider of industrial AI, demonstrated significant improvements at its Electronics Factory in Erlangen, Germany. They report that machine learning optimises testing procedures, significantly increasing first-pass yield (the percentage of products passing testing without rework)[133]. Higher first-pass yield directly translates to less rework and scrap, improving efficiency and throughput. Siemens also integrated AI in robotics: AI-enabled robots handle part picking and placement in fully automated assembly lines, which resulted in a 90% reduction in automation costs for those tasks[133]. At the same time, human operators use AI-guided systems for decision support, which has enhanced their productivity and quality of work[133]. Underpinning these deployments is Siemens' industrial-grade AI infrastructure of hardware/software that simplifies adoption and minimises change management [133]. They've set up automated training and deployment pipelines for AI models, reducing the effort to update models and monitor them continuously[133]. This infrastructure has helped foster trust and scalability, lowering barriers for rolling out AI solutions broadly. Siemens' case exemplifies a highly automated "smart factory" in action – AI is in quality control, in robotics, and in an enabling platform, yielding impressive efficiency and cost gains, and serving as a model for digital manufacturing.

Mengniu Dairy (Food and Beverage): Mengniu, one of the largest dairy companies, pursued AI integration across its supply chain and production. In their quality labs, they implemented AI modules like neural network image recognition and reinforcement learning-based intelligent scheduling to replace manual testing processes[133]. This ensured more accurate and efficient testing of dairy products. On the logistics side, Mengniu used AI to automate supplier order scheduling and delivery dispatch. The results were notable: inventory turnover increased by 73% and operational efficiency by 8% due to these AI optimisations[133]. In their factories, they deployed predictive maintenance to foresee equipment faults and avoid downtime[133]Improving overall production reliability. Mengniu's approach highlights AI not just in a single factory, but in connecting farm, factory, and distribution – a holistic supply chain improvement. It also shows AI's versatility, from lab testing (vision + RL) to procurement and logistics (machine learning optimisation) to maintenance.

Midea Group (Consumer Appliances): Midea, a global appliance manufacturer, undertook factory-wide AI adoption, applying AI in product design, production, quality, equipment management, energy management, and logistics[133]. One concrete outcome: they achieved a 25% reduction in product development cycles by using AI in design and development processes[133]. On the manufacturing side, they saw a 53% reduction in the rate of poor quality and a 29% optimisation of logistics paths through AI applications[133]. These are considerable improvements in both quality and supply chain efficiency. Notably, Midea managed to scale AI across 457 sub-scenarios in their factories, mainly by developing their small-sample AI algorithms and using open AI cloud platforms[133]. By focusing on algorithms that can learn from limited samples (important when data is scarce or when quick deployment is needed) and by having a cloud infrastructure, they reduced the time and cost of deploying AI at scale. Midea's success underlines the importance of scale and internal capability – they didn't just implement one or two use cases, but hundreds, creating a true digital manufacturing ecosystem that is continuously improving.

Each of these cases underscores various points. The importance of targeting high-impact use cases, the need for investing in people and infrastructure, and the tangible gains in efficiency, quality, and speed. From heavy industry to electronics to food, AI/ML is delivering real-world value. These examples also reflect that AI in manufacturing is maturing – it's not just lab experiments, but deployed solutions governing multimillion-dollar operations. They serve as inspiration and proof that, despite the challenges, AI can be successfully woven into manufacturing to achieve breakthrough performance.

Future Trends and Emerging Developments

Looking ahead, the role of AI/ML in manufacturing will continue to expand. Several future trends are on the horizon, poised to transform digital manufacturing further in the next decade:

Autonomous Factories and Lights-Out Manufacturing: We will see movement toward highly automated "lights-out" facilities that require minimal human presence on-site. With AI orchestrating production, machines can run unattended for extended periods – even in the dark, hence "lights-out." Already, some factories (in electronics and warehousing) have achieved this on specific shifts. AI will be the key to handling variability and decision-making in these environments, effectively becoming the plant manager on off shifts. Lights-out production uses AI, robotics, and IoT to automate every step of manufacturing, from raw materials to finished products[155]. In the future, expect more factories (especially for standardised products) to approach full autonomy, with human workers mostly supervising remotely or focusing on maintenance and improvements. A related concept is the "industrial metaverse" – a virtual representation of the factory (enabled by AI and AR/VR) where remote operators can monitor and control operations. Companies like Siemens speak of this vision, where AI and digital twins combine to allow factories to run themselves under virtual oversight(133), essentially. While not every factory will be lights-out (especially those with high mix or requiring craftsmanship), the technologies developed will raise automation levels across the board.

Federated Learning and Cross-Company Collaboration: Data sharing barriers have spurred interest in federated learning (FL) for manufacturing. In federated learning, multiple factories or companies collaborate to train a shared ML model without directly sharing their raw data. Each participant trains the model on their local data and shares only model updates (gradients), which are aggregated centrally to form a global model. This approach can help, for example, a group of suppliers or a consortium of manufacturers develop a more effective predictive maintenance model by leveraging each other's data on similar machines, all while keeping proprietary data in-house. Federated learning is an emerging concept of collaborative learning that can help smaller manufacturers learn from each other without exposing confidential data[156]. We foresee federated learning enabling industry-wide AI models (for quality, maintenance, yield optimisation) that are much more powerful than any single company's model, benefiting especially those who lack big data volume individually. Additionally, FL can be used within a large enterprise: multiple plants in a corporation can train models locally (accounting for local context) and still contribute to a unified global model. This not only protects data privacy but can reduce communication costs (since only model parameters, not complete datasets, are exchanged). As edge computing in factories grows, federated learning is a natural fit – essentially doing AI training at the edge collaboratively. We expect frameworks and standards to emerge to facilitate FL in industry. Over time, this can lead to "network effects" in manufacturing AI, as more participants join a federated learning network, all their models get better, accelerating industry-wide improvements.

Generative AI for Design, Simulation, and Operations: Generative AI will become more deeply integrated. In design engineering, generative design (as discussed with the GM example) will likely become commonplace for components, leveraging cloud computing to iterate thousands of designs that meet specified goals (weight reduction, strength, cost, etc.). This will shorten design cycles and yield innovative products. We'll also see AI-generated simulations: AI can create realistic simulation scenarios for testing production changes or new product introductions (almost like creating synthetic data to test "what-if" situations). For instance, before physically reconfiguring a production line, AI could generate a simulation of how the new layout would perform, including possible issues, which engineers can use to refine plans. In operations, large language models might be deployed as conversational assistants for factory personnel – an operator could ask an AI, "Why did we have a dip in output yesterday?" and the AI (connected to data and using natural language processing) could respond with a summary analysis. Generative AI could also help create on-the-fly work instructions or training content for workers, tailored to their level of expertise or preferred language. Another future use is in generative process control: given a desired outcome, the AI could generate the process recipe or parameters to achieve it (staying within known constraints). This flips the current paradigm – rather than tweaking knobs to see results, you tell the AI the result you want, and it proposes how to get there. While still early, these applications hint at an AI that not only optimises within set conditions but also creatively develops new solutions. We can also expect generative AI to aid in supply chain scenario planning, generating possible demand or disruption scenarios for planners to consider (some companies already use it to create synthetic demand scenarios to stress-test their plans[142]). The significant advances in AI models (GPT-like models, etc.) in recent years will be adapted to industrial contexts, albeit tuned for technical accuracy and reliability.

Digital Twins and Virtual Commissioning: The concept of digital twins – virtual replicas of physical assets or processes – will evolve further with AI. In the future, every critical machine or process could have a continuously AI-updated twin that not only mirrors its current state but also predicts its future state. This will be facilitated by IoT connectivity and advanced simulation models augmented by fundamental data. AI will enable twins to self-calibrate (keeping the simulation accurate as conditions change) and to run countless virtual experiments. For example, before ramping up a production line by 20%, an AI-driven twin could simulate the impact on each machine's load and quality output, revealing any bottlenecks or maintenance needed, thus de-risking changes. Virtual commissioning – testing and validating control logic or production setups in a virtual environment before implementing physically – will become standard, with AI assisting in automating those tests and even generating test cases (e.g., generative AI making up rare fault scenarios to test robustness). This trend will shorten deployment times for new production lines or products because much of the debugging happens in the virtual realm, and AI ensures the virtual tests are comprehensive. AI may also allow adaptive twins: not just static models, but ones that learn and improve in fidelity over time, eventually reaching a point where a significant portion of plant optimisation and troubleshooting can be done on the twin, saving downtime on the real equipment.

Human-Centric AI and Industry 5.0: There is a growing recognition that the next wave (sometimes dubbed Industry 5.0) will focus on human-centric technologies, bringing a balance between automation and human creativity. Future AI systems in manufacturing will likely emphasise collaboration, transparency, and ergonomics. This includes better human-machine interfaces: e.g., AR glasses that seamlessly overlay AI insights, voice-activated AI assistants on the line, and intuitive visualisation of AI data (like a quality heatmap an operator can view on a tablet). We'll also see more "explainable AI" by design, where AI tools used in operations come with built-in explainers so users quickly grasp the reasons behind suggestions. The role of the human will be reimagined – possibly termed as "augmentation engineers" or "AI-enabled operators", who manage fleets of AI agents. Decision-making will become a hybrid human-AI process: AI might narrow down options or provide a recommendation, and a human makes the final call using both the AI's input and their expertise. Over time, as trust increases, the balance may shift more towards AI for routine decisions and humans for strategic or ethical choices. This partnership could extend to maintenance (human + AI diagnosing together), quality (AI screens, human verifies edge cases), and even management (AI tracks KPIs and suggests improvements, humans set direction). Educational curriculums and training will adapt to produce this kind of workforce. The ultimate trend is ensuring that as AI grows more powerful, it remains aligned with human values and needs on the factory floor, enhancing jobs rather than making them devoid of meaning. In practice, this might mean designing AI systems with feedback loops where operators can correct or fine-tune AI behaviour easily (learning from human feedback), so the AI is constantly learning from humans and vice versa. The companies that succeed will likely be those that treat AI not as an infallible oracle, but as a collaborative team member whose "opinion" is valuable but works in concert with human judgment.

AI-Driven Custom Manufacturing and Lot-Size-One: Another future trend is hyper-personalised manufacturing – even at lot size one (each product custom-made) – enabled by AI. AI can manage the complexity of switching setups frequently, optimising sequences to minimise changeover time, and ensuring quality despite constant changes. This could allow economic production of individualised products (from sneakers with a custom design to cars tailored to each buyer's preferences) at near mass-production efficiency. AI's flexibility and learning ability will be key to such "mass customisation." This trend aligns with consumer demand for personalisation and could open new business models (manufacturing as a service, where small batches or one-offs can be profitable).

Continual Learning and Self-Healing Systems: Future manufacturing AI will likely incorporate more online learning models that continue to learn during operation. Right now, many deployed models are static until retrained offline. But with better algorithms, we'll see AI that adapts in real-time. For example, a robot's vision system might continuously improve its object detection as it considers more variations. Or a process control AI might learn from every cycle to get slightly better over time, effectively reaching optimal performance and staying there even as conditions evolve. Coupled with this is the idea of self-healing or self-optimising systems: if performance degrades (due to drift or anomaly), the AI detects it and takes corrective action or retrains itself to restore performance. This would reduce the need for human intervention in maintaining the AI models. It does introduce complexity (guarding against learning the wrong thing), but research into safe reinforcement learning and adaptive control is addressing that. Essentially, factories might eventually have AI that not only runs them but also keeps tuning itself up – a sort of autonomic manufacturing system.

In summary, the future of AI in manufacturing points to more autonomy, more collaboration (both human-AI and cross-organisation), and deeper integration from design to delivery. Factories will increasingly operate as smart, connected ecosystems with AI at their core – a vision of "smart factories" evolving into "cognitive factories" that can reason, adapt, and even learn on the fly. Manufacturers who stay abreast of these trends and experiment early will be well-positioned to capitalise on the next wave of digital transformation.

Best Practices for Implementing AI/ML in Manufacturing

For manufacturing organisations embarking on AI/ML initiatives, experience has shown that following certain best practices greatly improves the chances of success. Here are some key guidelines:

Start with Clear Goals and High-Impact Use Cases: Identify specific problems or opportunities where AI/ML can provide tangible value. Rather than adopting AI because it's trendy, focus on business-driven objectives – e.g., reducing predictive maintenance costs by 20%, or improving yield by 5%. Engage stakeholders to pinpoint pain points (excess scrap, frequent downtime, forecasting errors, etc.) and assess if AI/ML is the right tool to address them. Define precise success metrics (KPIs) for the project, such as defect rate reduction, downtime hours saved, or inventory turns improved[157]. Having clear goals helps maintain focus and allows you to measure ROI. Also, prioritise feasible use cases (data is available) and valuable (significant ROI) – a matrix scoring of potential projects on impact vs. ease can help choose a good starting project. Many recommend starting with a pilot project that is ambitious enough to matter but not so critical that failure is catastrophic – this provides a learning ground. Examining case studies from similar industries (as we did above) can inspire what's possible and realistic[157].

Ensure Data Readiness (Quantity, Quality, and Infrastructure): Before diving into modelling, invest in your data foundation. "Good data is at the heart of it all. Invest in data cleaning and preprocessing to eliminate inconsistencies and biases that could skew your AI model."[158]. Audit what data you have and what you need – you might need to install new sensors or start collecting data that wasn't recorded. Work on data integration early: break down silos and consolidate data from different sources (MES, ERP, sensors, etc.) into a unified repository or pipeline. Address data quality issues by cleaning historical data and setting up processes to clean new data continuously. This may involve removing erroneous entries, normalising units, aligning timestamps, and enriching data with context (labels, metadata). Establish a robust data pipeline for the AI project: how data flows from the factory floor (or database) to the model and back. Consider using data lake or warehouse solutions to store large datasets, and ensure you have the necessary tools (ETL software, IoT platform, etc.). If you have limited data, explore techniques like augmentation or starting with pre-trained models (transfer learning). And don't forget data governance – define who owns the data, access controls, and documentation so that the AI team and domain experts can collaborate on data easily. Essentially, treat data as a critical asset, because for AI, it is.

Build Multidisciplinary Teams (Combine Domain and Data Science Expertise): AI in manufacturing sits at the intersection of operations and analytics, so you need both domain experts and AI experts working together. Create a cross-functional team that includes manufacturing/process engineers, data scientists or ML engineers, IT/data architects, and, if possible, operators or technicians who know the day-to-day realities. The domain experts provide context: they help the data scientists understand what the data means (e.g., what is a "cycle complete" signal, or how a particular defect is defined) and validate whether model outputs make sense physically. The AI experts bring knowledge of algorithms, feature engineering, and so on. Encourage a culture of learning from each other – e.g., train engineers in basic data science and train data scientists in shop-floor operations. This also addresses the skills gap internally. Some companies have found success by embedding data scientists in plant teams or vice versa for a period, so they truly speak each other's language. If in-house skills are lacking, consider partnerships or hiring consultants, but ensure knowledge transfer to internal staff. Additionally, having an AI champion or product owner who understands both worlds and can translate between them helps keep everyone aligned. Regular communication and agile iteration (like weekly meetings to review intermediate results with engineers) can catch issues early (for example, noticing if the model is using a data input that is unreliable in real production). In short, don't silo the AI project in the IT department – integrate it with operations experts.

Start Small, Then Scale (Pilot Projects and Phased Rollout): Implement the chosen use case as a pilot in a controlled scope – for example, one production line, one cell, or one plant. This allows you to test the technology, demonstrate value, and learn lessons on a smaller scale. Keep the pilot's timeline reasonably short (a few months) to maintain momentum and show results. When deploying the pilot, do it in parallel or shadow mode first if possible (e.g., have the AI make predictions while operators still make decisions) to validate its performance without risking operations. Once the pilot meets success criteria, plan for scaling: how will you extend to other lines, products, or plants? Document the process and create templates or tools that can be reused. Also, anticipate how scaling affects infrastructure – maybe the pilot was done using a local server, but scaling to 10 plants might need a cloud setup, etc. Iterate based on pilot feedback: perhaps the model needs refinement or the user interface needs tweaks to be more user-friendly. An iterative, agile approach (develop -> test -> get user feedback -> refine) is very effective in AI projects. It's often said: think big, start small, scale fast. So have the big vision (smart factory, etc.) but nail a small piece of it first, then use that success to drive broader adoption and secure buy-in/funding for the next steps.

Integrate with Existing Workflows and Systems: AI solutions must fit into the operational workflow; otherwise, they won't be used or will cause disruption. Plan integration points with MES, ERP, SCADA, etc., early on. For example, if your ML model predicts machine failures, integrate it such that its output goes into the maintenance team's existing ticketing system or alerting mechanism – don't require them to check a separate AI dashboard that they might forget to look at. If AI provides a scheduling optimisation, feed that into the MES or scheduling tool the planners already use [158]. Essentially, make AI an enhancement of current systems, not a completely separate pathway. Use APIs or middleware to have systems talk to each other – modern software often can be integrated via REST APIs or message brokers. Where manual processes exist, consider how AI will be presented to users – perhaps via a simple web app or even email reports if that's what people are comfortable with initially. The goal is seamless integration: users might not even realise when they are interacting with AI versus traditional software. Also, ensure the AI system updates in sync with production – for instance, if a new product recipe is introduced, the AI model should be aware or retrained accordingly to incorporate it (linking with PLM could help here). A common mistake is treating the AI model deployment as the finish line; in truth, deployment is just the start. You need monitoring (to ensure the model is still performing well), maintenance (to update the model as data drifts or processes change), and alignment with any changes in upstream or downstream systems. By planning integration and maintenance from the get-go, you create an AI solution that's robust and lives on beyond the initial team that implemented it.

Address Change Management and Involve the Workforce: Proactively manage the human side of AI adoption. Communicate early and often with those who will be using or affected by the AI solution – explain what it does, how it will help, and importantly, that it's meant to assist, not replace them. Involve end-users in the design/testing phase: for example, get feedback from machine operators on an alert's format or from quality inspectors on an AI classification's usefulness. This inclusion not only improves the solution (making it user-friendly and practical) but also builds buy-in because people feel heard. Provide training programs for staff to learn about AI/ML basics relevant to their role. If an operator understands that a "prediction model" is like a more sophisticated alarm that looks at patterns, they might trust it more. Some companies run workshops or even simple cheat sheets to demystify AI. Leadership should champion the AI project to signal its importance, but also reassure that the workforce will be supported through the transition (no immediate job losses, perhaps re-skilling opportunities). It may help to highlight how AI will remove drudgery – e.g., "This vision system will save you from staring at products for 8 hours and free you to do higher-level quality analysis." Manage expectations as well – be honest that the AI might not be perfect at first and that their expertise is still vital to manage exceptions. Create feedback loops where workers can report AI errors or suggest improvements (and then iterate on the model). Recognise and reward employees who embrace and champion the new tools, creating positive reinforcement. Essentially, cultivate a company culture where people and AI work together, and continuous improvement is embraced. When fear or resistance is encountered, address it empathetically – perhaps by providing examples of other companies where AI adoption led to business growth and upskilled roles, or by pilot-testing with a willing group first and letting their success stories convince others. Change management is often the deciding factor between an AI pilot that dies and one that scales to enterprise-wide usage.

Focus on Governance, Ethics, and Security: From the outset, implement proper governance for your AI initiatives. This means setting up policies for data usage (ensuring compliance with any regulations like GDPR if personal data is involved, even if manufacturing data is not typically personal), model validation, and result verification. For critical decisions, have a human-in-the-loop or at least human oversight until the AI has proven itself over time. Document the AI model's development process and rationale, especially if needed for compliance (e.g., traceability of how a quality decision was made). Keep an eye on biases – even in manufacturing, biases can creep in (for example, an AI might systematically under-predict failures on a particular older machine type because most data came from newer machines, effectively biasing against that scenario). Test the model on various scenarios to ensure it's robust and fair. Maintain transparency with employees and perhaps even customers (e.g., if AI is used in inspection, customers might be interested in the fact that every product is AI-checked for defects – this could be a marketing point if framed correctly). On the cybersecurity front, integrate the AI system into your IT security audits. Secure data pipelines (encrypted data transfer, access control to models). Since AI might open new connections (like devices sending data to the cloud), work with cybersecurity teams to mitigate risks (VPNs, firewalls, anomaly detection for your IIoT network). Also consider fail-safes: if the AI system fails or gives wrong output, make sure the default state is safe – e.g., an AI control should fail back to manual or a secure mode. Having a clear governance structure – perhaps an AI steering committee or appointing someone as AI lead – ensures ongoing oversight. This group can also set guidelines like "We will not use AI to monitor individual worker productivity in a punitive way" to alleviate ethical concerns and ensure AI aligns with company values. By treating governance and ethics not as red tape but as essential design parameters, you build trust internally and externally and avoid pitfalls that could arise from misuse or unintended consequences of AI.

Monitor, Measure, and Iterate: Once the AI solution is deployed, continuously monitor its performance against the defined metrics. Put in place a mechanism to catch model drift or degrading accuracy – for instance, schedule periodic evaluation of predictions vs. actual outcomes. In manufacturing, processes can change (different product mix, new machines, etc.), and models might need retraining. So, plan for an update cycle: who is responsible for retraining the model when required? Do you need to collect more labelled data over time to improve it? Treat the AI model as a living system rather than a one-time solution. Additionally, measure the business impact – did it achieve the ROI expected? If not, why does the model need improvement, or was the target unrealistic, or is adoption lacking? Use these insights to iterate both on the technical solution and on how it's used. Many companies do a post-mortem or retrospective after a pilot, capturing lessons learned in data, integration, user training, etc., to apply to the next project. Celebrate successes and publicise them internally (and even externally as appropriate) – for example, if AI reduced scrap by X%, let everyone know; it reinforces the value of AI and creates momentum for further projects. Conversely, be honest about shortcomings and treat them as learning opportunities to refine the approach. Over time, as you roll out more AI use cases, establish a knowledge base or playbook – a set of best practices (which variables typically matter for quality predictions? what pitfalls to avoid in sensor data integration? etc.) that can be reused. This helps scale AI capability across the organisation without reinventing the wheel each time. Essentially, adopt a continuous improvement mindset for AI projects, just as one would for manufacturing processes themselves (aligning well with Lean/Agile philosophies).

By following these best practices, from strategic planning and data groundwork to people-centric deployment and iterative refinement, manufacturers can significantly increase the likelihood that their AI/ML initiatives will deliver meaningful, sustained benefits. Implementing AI in manufacturing is a journey that blends technology with process and culture. With careful navigation using the guidelines above, companies can transform that journey from a risky experiment into a well-charted path toward the next level of operational excellence.

Conclusion and Future Outlook in Digital Manufacturing

Digital manufacturing represents the transformative integration of advanced digital technologies into product design and production. It encompasses the use of information and communication technologies across the entire factory ecosystem to improve efficiency, agility, and quality. In practice, this means linking design, engineering, and production through networked systems and data flows. Nearly all industry leaders now recognise that digital manufacturing (often called "smart manufacturing") is essential for competitiveness; for example, a recent survey found that 92% of manufacturers see smart manufacturing as the main driver of competitiveness over the next several years. www2.deloitte.com. In summary, digital manufacturing as a concept sets the stage for the rest of this book by showing how connected, data-driven production systems enable faster innovation, more customised products, and leaner operations without getting bogged down in technical detail.

Cyber-Physical Systems (CPS) are the bedrock of smart factories. They tightly integrate physical machinery (sensors, actuators, and controls) with digital computing and networking. In a CPS-enabled shop floor, machines and sensors continuously collect data, feed it into computational algorithms, and then use the results to control and optimise the physical processes in real time. As defined by the U.S. National Science Foundation, a CPS "integrates sensing, computation, control and networking into physical objects and infrastructure". This fusion of hardware and software makes it possible to monitor production parameters instantaneously and to adjust machines for optimal performance automatically. Strategically, CPS bring agility and resilience: they allow manufacturers to detect faults early, adjust production schedules on the fly, and maintain continuous quality control.

In short, CPS is what makes "Industry 4.0" a reality by enabling the real-time visibility and control that all other digital manufacturing technologies rely upon.

Cloud and edge computing are complementary pillars of modern manufacturing IT architecture. Cloud computing provides virtually unlimited storage and processing power in centralised data centres, enabling powerful analytics, collaboration, and global integration. By hosting simulations, analytics platforms, and enterprise resource planning in the cloud, manufacturers can scale resources up or down as needed and share data across locations. Edge computing, by contrast, brings intelligence closer to the shop floor. Edge devices (local servers or intelligent gateways) collect machine data and run analytics at or near the source to deliver millisecond response times and reduce dependence on continuous internet connectivity. This hybrid approach is now mainstream: many manufacturers prioritise sensors, cloud, and edge computing as core investments. Together, cloud and edge systems give manufacturing leaders the best of both worlds – robust big-data analysis and AI in the cloud, plus reliable real-time control and decision-making at the edge. Their strategic importance lies in enabling data-driven automation and enterprise-wide visibility without compromising responsiveness or security.

The concept of the digital twin is a natural extension of CPS and data analytics. A digital twin is a high-fidelity, virtual replica of a physical asset, process, or entire factory. It continuously ingests live data from sensors, control systems, and other sources to mirror the real system's status. As McKinsey describes it, digital twins are "real-time virtual renderings of the physical world". With digital twins, engineers can run simulations and "what-if" scenarios on an exact model of the production line or product. This allows teams to optimise layouts, predict bottlenecks, and evaluate changes without risking actual downtime. In practice, twins exist at many scales: from individual products (tracking as-built conditions for quality and warranty analysis) to factory-floor simulations (optimising throughput and maintenance). Adoption has skyrocketed – a recent executive survey found that 86% of manufacturers see a digital twin as applicable to their business, and nearly half have already deployed one. The strategic payoff is clear: digital twins unlock better planning, preventive maintenance, and continuous improvement by letting manufacturers experiment in the virtual world before making costly physical changes.

A Manufacturing Execution System (MES) is the software backbone on the shop floor that translates business plans into action. It acts as a bridge between high-level planning (like ERP) and the plant equipment. In effect, an MES "monitors and controls production processes on the shop floor" by collecting real-time data from machines, operators, and materials. The MES tracks every step of turning raw materials into finished goods, providing visibility into production status, inventory, and quality metrics. By enforcing standard processes and capturing live data, MES software lets managers spot bottlenecks and variances as they happen. It also automates work orders and dispatches instructions to operators, ensuring on-time scheduling and optimal resource use. This results in higher throughput and lower downtime. For example, MES systems provide real-time visibility that enables dynamic scheduling and immediate adjustments to minimise delays. In summary, MES is strategically crucial because it makes the factory floor transparent and controllable, ensuring that digital plans translate into efficient, high-quality production. Indeed, many manufacturers now list MES or advanced production scheduling among their top technology investments.

Product Lifecycle Management (PLM) systems cover the entire lifespan of a product, from initial concept through design, production, and service. Traditionally seen as engineering data management, PLM has evolved into a central repository and process framework for end-to-end product information. Modern PLM integrates CAD models, bill of materials, change requests, compliance documentation and more, breaking down silos between departments. This holistic view – often realised through the vision of a "digital thread" – ensures that manufacturing always has up-to-date product data. In essence, PLM provides the infrastructure that ties product development and manufacturing processes together. As one industry analysis noted, PLM is now viewed as "an indispensable infrastructure driving end-to-end product development lifecycles", manufacturing-today.com. Leaders are moving PLM systems to the cloud and enhancing them with AI and augmented reality to facilitate collaboration and innovation. Strategically, a robust PLM solution allows companies to accelerate new product introduction, improve cross-functional collaboration, and maintain data consistency, all of which translate to faster time-to-market and better product quality.

Robotics and automation remain at the heart of the productivity gains in modern factories. Across all industries, the deployment of industrial robots has grown to record levels. For instance, over 3.5 million robots were in operation worldwide as of 2021, and demand for new installations is at an all-time high. Automation ranges from traditional robotic arms that perform welding and assembly to autonomous guided vehicles (AGVs) that handle material transport. A newer trend is collaborative robots, or "cobots", designed to work safely alongside humans. Coupled with improved AI vision systems and machine learning, robots today are more flexible and easier to reprogram, allowing plants to switch production lines quickly and perform complex tasks. The strategic impact is profound: robotics enables factories to operate around the clock, improve precision, and alleviate labour shortages. Moreover, as the industry shifts toward mass customisation, advanced automation means that even small-batch or unique products can be manufactured efficiently. In summary, robotics and automation technologies provide the muscle of digital manufacturing – they turn data and plans into physical action on the factory floor.

Data is the lifeblood of digital manufacturing. From sensor streams on machines to records in PLM and MES, the volume of information generated by modern plants is enormous. Harnessing this data—through collection, integration, and analysis—is what unlocks the value of all other technologies. Companies today are acutely aware of this: many are prioritising the establishment of robust data infrastructures so that they can apply analytics and AI. Clean, well-governed data enables predictive maintenance (by spotting equipment anomalies early), quality control (by correlating process parameters to defects), and supply-chain optimisation (through real-time tracking of parts). In a recent Deloitte study, 78% of manufacturers reported allocating over 20% of their improvement budget toward building these smart manufacturing and data foundations. Still, many firms struggle with data silos and quality. Nearly three-quarters of manufacturers in one survey said they are increasing investment in data lifecycle management to prepare for AI, yet 70% cite data quality and integration issues as the most significant obstacles. Thus, strategically, mastering data – making it accessible, accurate and secure – is the key enabler for all digital manufacturing innovations. When done correctly, data-driven insights give decision-makers the visibility needed to optimise operations continuously.

Machine learning (ML) and artificial intelligence (AI) are the decision engines of the modern factory. While all the afore mentioned technologies generate data, AI transforms that data into smart actions. Applications are widespread: ML algorithms can predict machine failures before they happen, optimise production schedules dynamically, and sort parts by quality using computer vision. Generative AI is also making inroads – not only in automating back-office tasks but in engineering design and customer service. For example, many manufacturers are already using generative AI chatbots to help customers choose product features, or using AI to propose new design variants based on historical product data. A recent Deloitte survey found that 55% of industrial manufacturers are already leveraging generative AI tools in some form, and over 40% plan to increase investment in AI and ML in the next few years. These technologies are proving their value: manufacturers report that AI and ML have among the highest impact on business outcomes of any smart factory tech, improving efficiency, productivity, and cost reduction. Strategically, AI/ML tie together the entire ecosystem – they make the MES smarter, the digital twin more predictive, and the supply chain more resilient. However, their success depends on good data and clear business cases, so leading companies are carefully targeting AI investments to high-ROI use cases.

Looking Ahead: Future Trends in Digital Manufacturing

The convergence of these technologies points toward ever more autonomous and adaptive factories in the next 5–10 years. One emerging vision is the autonomous factory, where AI-driven systems and robotics continuously self-optimise production. In such a setting, machines would adjust speeds, reconfigure workflows, and order parts from suppliers on their own, all while meeting tight quality and energy-use targets. Generative AI will play a growing role: not only in design and customer experience, but also in synthesising insights from vast engineering archives to spur innovation on existing products. Sustainability will also reshape digital manufacturing. Advanced simulations and digital twins will be used to minimise waste and energy consumption – even a 1% efficiency gain in industry can save nearly 90 million tons of CO_2 annually. Manufacturers will increasingly measure and reduce their environmental footprint through data-driven optimisation. At the same time, the workforce will continue to evolve: digital skills, AI literacy, and systems integration know-how will become table stakes. Companies will invest heavily in training and collaboration tools (including AR/VR for immersive training) to upskill their teams, recognising that human–machine collaboration (sometimes called "Industry 5.0") offers a competitive advantage.

Other future directions include further virtualisation of supply chains (using IoT and blockchain for end-to-end visibility) and the spread of additive manufacturing for custom, on-demand production. Cybersecurity will also be an ever-present concern as connectivity grows. In sum, the factory of tomorrow will be vastly more intelligent and resilient, but realising that vision requires leadership and action today. As a call to action, industry leaders should continue to invest in integration (ensuring systems and data work together), cultivate talent (developing digital skills), and embrace innovation (piloting AI and digital twins) to stay ahead. By drawing on the strategies outlined throughout this book, manufacturers can turn digital tools into durable competitive strengths.

Overall, digital manufacturing is not a destination but an ongoing journey of transformation. The key insight from this book is that success depends as much on strategy and culture as on technology. As we move forward, companies that adopt a forward-looking mindset – combining cutting-edge tools like generative AI with sustainable practices and skilled people – will lead the industry into its next era. The future factory may be more automated and data-driven than ever, but it will be guided by human innovation and responsibility.

Digital manufacturing weaves together technologies from CPS to AI to create agile, efficient production systems. Each chapter's theme (from CPS networking to PLM integration) plays a vital role. Together, they form a blueprint for continuous improvement. Looking ahead, the emphasis will be on intelligent automation, sustainability, and human–machine synergy. Leaders who tie these advances to clear business goals and workforce development will be best positioned to thrive in the next decade of manufacturing.

References

1.   Sirimanne S. What is 'Industry 4.0' and what will it mean for developing   countries?   [Available   from: https://www.weforum.org/stories/2022/04/what-is-industry-4-0-and-could-developing-countries-get-left-behind/#:~:text=,towards%20more%20technologically%20advanced%20sectors.
2.   Matej Bajgar SC, Flavio Calvino, Chiara Criscuolo, Jonathan Timmis. Bits and bolts: The digital transformation and manufacturing. OECD Science, Technology and Industry Working Papers. 2019.
3.   Revolution FI. Fourth Industrial Revolution   [Available from: https://www.weforum.org/focus/fourth-industrial-revolution/#:~:text=The%20Fourth%20Industrial%20Revolution%20represents,opportunity%20to%20help%20everyone%2C%20including.
4.   Symestic.   Digital   Manufacturing   [Available   from: https://www.symestic.com/en-us/what-is/digital-manufacturing#:~:text=,bridging%20physical%20and%20digital%20domains.
5.   Manuel Schuler WK. Emerging technology trends shaping the digital   manufacturing   landscape   [Available   from: https://www.bearingpoint.com/en-us/insights-events/insights/emerging-technology-trends-shaping-the-digital-manufacturing-landscape/#:~:text=Artificial%20Intelligence%20,environment%20and%20autonomously%20executing%20tasks.
6.   Room IP. Global Robot Density in Factories Doubled in Seven Years   [Available from: https://ifr.org/ifr-press-releases/news/global-robot-density-in-factories-doubled-in-seven-years#:~:text=Frankfurt%2C%20Nov%2020%2C%202024%20%E2%80%94,IFR.
7.   Verzelen F. Are virtual twins key to achieving world's sustainability   goals?   [Available   from: https://www.weforum.org/stories/2021/01/virtual-twin-technology-sustainability-

goals/#:~:text=from%20five%20industries%3A%20construction%2C%2
0consumer,3%20trillion%20of%20economic%20value.

8.	Components V. An Introduction to Digital Manufacturing [Available from:	https://www.visualcomponents.com/blog/an-introduction-to-digital-manufacturing/.

9.	Tim Gaus MS. 2025 Smart Manufacturing and Operations Survey: Navigating challenges to implementation	[Available from: https://www2.deloitte.com/us/en/insights/industry/manufacturing/2025-smart-manufacturing-survey.html.

10.	Ewelina Gregolinska RK, Frédéric Lefort, Prashanth Parthasarathy. Capturing the true value of Industry 4.0 [Available from: https://www.mckinsey.com/capabilities/operations/our-insights/capturing-the-true-value-of-industry-four-point-zero.

11.	Manufacturing Leadership Council. Smart Factories and Digital Production	survey	[Available	from: https://manufacturingleadershipcouncil.com/how-close-is-the-smart-factory-of-the-future-36125/?stream=all-news-insights#:~:text=How%20widespread%20are%20digital%20factories%
3F,digitized%20their%20factory%20operations%20extensively.

12.	Business P. Digital Manufacturing Market	[Available from: https://www.precisionbusinessinsights.com/market-reports/global-digital-manufacturing-market#:~:text=Market%20Size%20in%202023.

13.	Reinhard Geissbauer JV, Stefan Schrauf Industry 4.0: Building the	digital	enterprise	[Available	from: https://www.pwc.com/gx/en/industries/industries-4.0/landing-page/industry-4.0-building-your-digital-enterprise-april-2016.pdf#:~:text=3,additional%20digital%20revenues%20and%20opti
mising.

14.	Kaplan I. Why manufacturers should use digital twins for sustainability	not	just	productivity	[Available	from: https://www.weforum.org/stories/2023/05/digital-twins-manufacturing-sustainability/#:~:text=sustainability,cost%20savings%20and%20impro
ve%20efficiency.

15.	Ibrahim Abdullah Alshunaifi CT. This is how we upskill future leaders to drive the digital transformation of manufacturing [Available from:	https://www.weforum.org/stories/2025/04/this-is-how-we-upskill-future-leaders-to-drive-the-digital-transformation-of-manufacturing/#:~:text=demand%20for%20a%20workforce%20skilled,
capable%20of%20navigating%20this%20shift.

16.	Council ML. How Close Is the Smart Factory of the Future? [Available from: https://manufacturingleadershipcouncil.com/how-close-is-the-smart-factory-of-the-future-36125/?stream=all-news-insights.

17. Dataparc. Understanding Big Data in Manufacturing [Available from: https://www.dataparc.com/blog/the-role-and-importance-of-big-data-in-manufacturing/#:~:text=Big%20data%20in%20manufacturing%20refers,as%20audio%2C%20video%2C%20or%20text.

18. Lauzier J. Big Data in Manufacturing: Unlocking Valuable Use Cases [Available from: https://www.machinemetrics.com/blog/big-data-in-manufacturing#:~:text=can%E2%80%99t%20be%20effectively%20processed%20using,that%20goes%20into%20modern%20manufacturing.

19. Paul Maplesden. Learn top types of tech for collecting manufacturing data [Available from: https://www.techtarget.com/searcherp/tip/Learn-how-manufacturing-data-is-collected#:~:text=IoT,due%20to%20the%20following%20factors.

20. intone. Challenges and Solutions in Real-time Data Management in Manufacturing Operations [Available from: https://intone.com/data-management-challenges-and-solutions-in-manufacturing-operations/#:~:text=Volume%20and%20Velocity%20of%20Data%3A.

21. Precognize. Data Historian [Available from: https://precog.co/glossary/data-historian/#:~:text=A%20data%20historian%20automates%20the,draw%20on%20it%20when%20needed.

22. Emqx. Introduction to MQTT Publish-Subscribe Pattern [Available from: https://www.emqx.com/en/blog/mqtt-5-introduction-to-publish-subscribe-model.

23. Dataparc. Advanced Analytics in Manufacturing: How It Compares to Operational Analytics [Available from: https://www.dataparc.com/blog/advanced-analytics-vs-operational-analytics-in-manufacturing/#:~:text=It%20is%20common%20to%20describe,four%20separate%20types%20of%20analytics.

24. SME. Unlocking Operational Efficiency: The Comprehensive Benefits of Predictive Maintenance in Manufacturing [Available from: https://www.sustainablemanufacturingexpo.com/en/articles/operational-efficiency-predictive-maintenance.html#:~:text=One%20of%20the%20most%20significant,the%20impact%20on%20production%20schedules.

25. Lee S. 5 Data Monetization Methods Transforming Your Manufacturing ROI [Available from: https://www.numberanalytics.com/blog/application-data-monetization-manufacturing-

strategies#:~:text=According%20to%20McKinsey%2C%20manufactur
ers%20implementing,2.
26. Shoplogix. Federated Learning in Manufacturing: The Power of
Distributed Data Explained [Available from:
https://shoplogix.com/federated-learning-in-
manufacturing/#:~:text=Basically%2C%20federated%20learning%20is
%20a,updates%20derived%20from%20that%20data.
27. Intraratio. What You Need to Know About Data Traceability in
Manufacturing [Available from: https://blog.intraratio.com/what-you-
need-to-know-about-data-traceability-in-
manufacturing#:~:text=Data%20traceability%20refers%20to%20the,be
%20easily%20recalled%20if%20necessary.
28. Emqx. Incorporating the Unified Namespace with ISA-95: Best
Practices [Available from:
https://www.emqx.com/en/blog/incorporating-the-unified-namespace-
with-isa-95-best-practices.
29. Sean Camarella MPC, Kevin Goering, Mark Huntington. Digital
twins: The next frontier of factory optimization [Available from:
https://www.mckinsey.com/capabilities/operations/our-insights/digital-
twins-the-next-frontier-of-factory-optimization.
30. Shoplogix. Interoperability Problems in Smart Factories: Bridging
the Gap in Industry 4.0 [Available from:
https://shoplogix.com/interoperability-problems-in-smart-
factories/#:~:text=Interoperability%20Problems%20in%20Smart%20Fa
ctories%3A,connection%20point%20between%20systems.
31. Kuuse M. Manufacturing Analytics – A Simple Guide for SMEs
[Available from: https://www.mrpeasy.com/blog/manufacturing-
analytics/#:~:text=,in%20manufacturing%20analytics%20is%20to.
32. Hoppe S. A global standard for industrial interoperability with
common data models [Available from: https://www.isa.org/intech-
home/2019/november-december/features/opc-ua-the-united-nations-
of-
automation#:~:text=United%20Nations%20of%20automation.%20,colla
boration%20partners%20across%20the%20world.
33. Hivemq. Smart Manufacturing Using ISA95, MQTT Sparkplug
and the Unified Namespace [Available from:
https://www.hivemq.com/resources/smart-manufacturing-using-isa95-
mqtt-sparkplug-and-uns/.
34. Cawsey M. A Data Monetization Strategy - Get More Value from
Your Master Data.
35. Leeson T. The Reach for Data Monetization.

36. Kannan R. The future of manufacturing: Self-optimizing factories powered by IIoT and digital twins.

37. Amplifi. Manufacturing a data culture: 6 tips to implementing Data Governance.

38. Cerexio. A Startup Guide to Deploy Robotic Machinery in Manufacturing Facilities [Available from: https://cerexio.com/blog/startupguide-deploy-robot-automation-facturies#:~:text=Industrial%20robots%20are%20typically%20used,The%20report%20records%20that%2C%20as.

39. Solutions P. What are the Different Types of Industrial Robots and Their Applications? [Available from: https://processsolutions.com/what-are-the-different-types-of-industrial-robots-and-their-applications/.

40. Automation W. AGVs vs AMRs: What's the difference and which one is right for you? [Available from: https://www.mwes.com/agvs-vs-amrs-whats-the-difference-and-which-one-is-right-for-you/#:~:text=AGVs%20work%20best%20in%20a,term%20commitment%20to%20the%20infrastructure.

41. Room IP. Collaborative Robots - How Robots Work alongside Humans [Available from: https://ifr.org/ifr-press-releases/news/how-robots-work-alongside-humans#:~:text=Cobots%20offer%20a%20quick%20entry,production%20needs%20are%20constantly%20changing.

42. Automation AfA. Revolution on the Factory Floor: How Automation is Redefining the Future of Manufacturing [Available from: https://www.automate.org/news/-86#:~:text=Computer%20and%20electronics%20technologies%20gave,flexible%20while%20attaining%20greater%20efficiency.

43. Latzel M. Challenges & Benefits of Robotic Manufacturing [Available from: https://www.essert.com/blog/robotics/challenges-benefits-robotic-manufacturing/#:~:text=The%20initial%20investment%20in%20automated,2.

44. Robotnik. Advantages and benefits of industrial robots: improving the 4 D's.

45. Rheinland T. Industrial Robotics and Cybersecurity: How Manufacturers Can Minimize Risk and Ensure Safe Operation [Available from: https://www.aem.org/news/industrial-robotics-and-cybersecurity-how-manufacturers-can-minimize-risk-and-ensure-safe-operation#:~:text=The%20need%20for%20safety%20in,and%20possibly%20effect%20physical%20damage.

46. Excelpro. The automation pyramid.

47. Müller C. Industrial Robots [Available from: https://ifr.org/img/worldrobotics/Executive_Summary_WR_Industrial_R obots_2022.pdf#:~:text=,3.

48. Greenawalt T. Amazon has more than 750,000 robots that sort, lift, and carry packages—see them in action [Available from: https://www.aboutamazon.com/news/operations/amazon-robotics-robots-fulfillment-center.

49. Automation AfA. Collaborative Robots (Cobots) Are Revolutionizing Flexible Manufacturing [Available from: https://www.automate.org/robotics/blogs/cobots-revolutionizing-flexible-manufacturing#:~:text=ImageThe%20demand%20for%20flexible%2C %20productive%2C,and%20agility%20to%20stay%20competitive.

50. Geeksforgeeks. What is Cloud Robotics? [Available from: https://www.geeksforgeeks.org/what-is-cloud-robotics/.

51. Hoppe S. There Is No Industrie 4.0 without OPC UA [Available from: https://opcconnect.opcfoundation.org/2017/06/there-is-no-industrie-4-0-without-opc-ua/#:~:text=meaning,it%20in%20a%20targeted%20and.

52. Dirjish M. Securing Safe Robotic Workplaces [Available from: https://www.automationworld.com/factory/safety/article/55253353/ensu ring-safe-robotic-workplaces-in-manufacturing.

53. Foundation UNS. Cyber-Physical Systems (CPS) [Available from: https://www.nsf.gov/funding/opportunities/cps-cyber-physical-systems#:~:text=Cyber,agriculture%2C%20aeronautics%2C%20buildi ng%20design%2C%20civil.

54. Arcot RV. Cyber-Physical Systems: The Core of Industry 4.0 [Available from: https://blog.isa.org/cyber-physical-systems-the-core-of-industry-4.0#:~:text=According%20to%20the%20NIST%20Special,multiple%20 economic%20sectors%20in%20the.

55. Raza M. Cyber-Physical Systems (CPS) Explained [Available from: https://www.splunk.com/en_us/blog/learn/cyber-physical-systems.html#:~:text=What%20is%20a%20cyber.

56. HTN. Cyber-Physical Systems (CPS) and Supply Chain Management [Available from: https://htn-expert.com/blog/cps#:~:text=example%2C%20sensors%20and%20sm art%20devices,they%20can%20quickly%20reconfigure%20their.

57. Salgarkar R. Cyber-Physical Systems (CPS) Industry [Available from: https://www.marketsandmarkets.com/PressReleases/cyber-physical-systems-cps.asp#:~:text=Higher%20growth%20in%20industrial%20automation, maintenance%20and%20optimization%20of%20resources.

58.   Defendere D. Top Applications of Cyber-Physical Systems in Manufacturing: Today and Shaping the Future   [Available from: https://datadefendere.medium.com/top-applications-of-cyber-physical-systems-in-manufacturing-today-and-shaping-the-future-99e3c4011440.

59.   Symestic. Cyber-Physical Systems (CPS)   [Available from: https://www.symestic.com/en-us/what-is/cps#:~:text=,physical%20processes%20to%20prevent%20accidents.

60.   Banerjee S, Jesubalan NG, Kulkarni A, Agarwal A, Rathore AS. Developing cyber-physical system and digital twin for smart manufacturing: Methodology and case study of continuous clarification. Journal of Industrial Information Integration. 2024;38:100577.

61.   Abinaya P, Kumar SP, Sivaprakash P, Kumar KA, Shuriya B, Surbhi Bhatia K. Smart Manufacturing with a Digital Twin–Driven Cyber‐physical System: Case Study and Application Scenario. Digital Twins in Industrial Production and Smart Manufacturing: An Understanding of Principles, Enhancers, and Obstacles: IEEE; 2024. p. 247-76.

62.   Stouffer KA. NIST Releases Cybersecurity Guide for Manufacturing Control Systems   [Available from: https://www.nist.gov/news-events/news/2022/03/nist-releases-cybersecurity-guide-manufacturing-control-systems#:~:text=NIST%20Releases%20Cybersecurity%20Guide%20for,lines%2C%20and%20other%20physical.

63.   Oberoi S. Cyber physical systems – Need for standardization [Available from: https://jtc1info.org/cyber-physical-systems-need-for-standardization/#:~:text=Cyber%20physical%20systems%20%E2%80%93%20Need,modelled%20in%20its%20abstract.

64.   ISO. ISO/IEC AWI TS 5689.

65.   NIST. Framework for Cyber-Physical Systems: Volume 1, Overview   [Available   from: https://nvlpubs.nist.gov/nistpubs/SpecialPublications/NIST.SP.1500-201.pdf#:~:text=%5BPDF%5D%20Framework%20for%20Cyber,for%20shaping%20the%20approach.

66.   Hat R. Understanding edge computing for manufacturing [Available   from: https://www.redhat.com/en/topics/edge-computing/manufacturing#:~:text=Edge%20computing%20%20is%20a,and%20more%20timely%20data%20analysis.

67.   Fogwing. Why Cloud Computing is Essential for the Modern Manufacturing   Industry?   [Available   from: https://www.manufacturingtomorrow.com/story/2023/01/why-cloud-

computing-is-essential-for-the-modern-manufacturing-industry/19950/#:~:text=Cloud%20computing%20provides%20services%20to,applications%20and%20other%20advanced%20technologies.

68.   Immerman G. Edge Computing vs. Cloud Computing in Manufacturing   [Available   from: https://www.machinemetrics.com/blog/edge-computing-vs-cloud-computing#:~:text=In%20the%20cloud%2C%20all%20data,reliable%20remote%20access%20to%20data.

69.   Lee S. 5 Proven Edge Computing Strategies to Boost Manufacturing   [Available   from: https://www.numberanalytics.com/blog/edge-computing-manufacturing-5-strategies#:~:text=Manufacturing%20industries%20often%20face%20strict,minimizing%20exposure%20and%20transmission%20risks.

70.   Miller D. How BMW Improves Interoperability with OPC UA and Edge   Computing   [Available   from: https://www.automationworld.com/communication/article/21521520/bmw-opc-ua-and-edge-computing-use.

71.   Team DA. Innovating Aerospace Manufacturing, Maintenance, and   Mobility   with   Cloud   Edge   Solutions   [Available   from: https://www.aviationtoday.com/2021/05/17/opinion-innovating-aerospace-manufacturing-maintenance-mobility-cloud-edge-solutions/.

72.   Moody PE. How manufacturing IT drives growth for Polamer Precision   [Available   from: https://www.ame.org/target/articles/2018/how-manufacturing-it-drives-growth-polamer-precision#:~:text=JobBOSS%20for%20Master%20Scheduling,%E2%80%9D.

73.   Ouyang C. Edge computing and hybrid cloud: scaling AI within manufacturing   [Available   from: https://www.ibm.com/products/blog/maximizing-enterprise-value-ai-edge-computing#:~:text=We%20manage%20the%20AI%20models,save%20millions%20of%20dollars%20annually.

74.   Burke T. How TSN helps manufacturers enable edge computing capabilities [Available from: https://www.plantengineering.com/how-tsn-helps-manufacturers-enable-edge-computing-capabilities/#:~:text=Time,Using%20TSN%20lets.

75.   Kayvaun Rowshankish RWZ, and Tomás Lajous What is digital-twin technology? [Available from: https://www.mckinsey.com/featured-insights/mckinsey-explainers/what-is-digital-twin-technology.

76.   Portley J. 6 Ways Digital Twins Are Revolutionising Manufacturing in 2024 [Available from: https://knowhow.distrelec.com/internet-of-things/6-ways-digital-twins-are-revolutionising-manufacturing-in-2024/#:~:text=A%20digital%20twin%20is%20a,performance%20within%20a%20virtual%20space.
77.   Airbus. Digital Twins: Accelerating aerospace innovation from design to operations [Available from: https://www.airbus.com/en/newsroom/stories/2025-04-digital-twins-accelerating-aerospace-innovation-from-design-to-operations#:~:text=Central%20to%20this%20vision%20are,world%20counterpart.
78.   IBM. What is a digital twin? [Available from: https://www.ibm.com/think/topics/what-is-a-digital-twin#:~:text=A%20digital%20twin%20is%20a,reasoning%20to%20help%20make%20decisions.
79.   Belden. Digital Twins and Edge Computing Unite to Drive Manufacturing Insights [Available from: https://www.belden.com/blog/industrial-automation/2024/05/28/digital-twins-and-edge-computing-unite-to-drive-manufacturing-insights#:~:text=Digital%20Twins%20and%20Edge%20Computing,latency%20since%20data%20is.
80.   Brewer A. Revolutionizing Manufacturing with Edge Computing and Digital Twins [Available from: https://medium.com/@alex.brewer/revolutionizing-manufacturing-with-edge-computing-and-digital-twins-02ebbbd43850.
81.   Shao G, Kibira D, Frechette SP. Digital Twins for Advanced Manufacturing: The Standardized Approach. Digital Twins, Simulation, and Metaverse, Springer Nature, Washington, DC; 2024.
82.   Mendi AF. A Digital Twin Case Study on Automotive Production Line. Sensors (Basel). 2022;22(18).
83.   Lee S. 7 Data-Driven Insights on Digital Twin in Manufacturing [Available from: https://www.numberanalytics.com/blog/digital-twin-manufacturing-insights#:~:text=7%20Data,35%25.
84.   EDI. Digital Twins in Automotive Manufacturing: Dynamic Virtual Models Advantages [
85.   Hasan M. Digital twin market: Analyzing growth and emerging trends [Available from: https://iot-analytics.com/digital-twin-market-analyzing-growth-emerging-trends/#:~:text=Trend%201%3A%20Digital%20twins%20deployed,to%20meet%20sustainability%20goals.

86. Sean Olcott CM. Digital Twin Consortium Defines Digital Twin [Available from: https://www.digitaltwinconsortium.org/2020/12/digital-twin-consortium-defines-digital-twin/#:~:text=Digital%20Twin%20Consortium%20Defines%20Digital,a%20specified%20frequency%20and%20fidelity.

87. Scoltock J. How digital twins are transforming aerospace development and testing [Available from: https://www.aerospacetestinginternational.com/features/how-digital-twins-are-transforming-aerospace-development-and-testing.html#:~:text=How%20digital%20twins%20are%20transforming,from%20aircraft%20development%20to%20operations.

88. Max R. What Are The Use Cases of Digital Twin in Aerospace? [

89. Jones G. Digital Twin Technology – The Smart Approach to Enhancing Manufacturing Productivity [Available from: https://metrology.news/digital-twin-technology-the-smart-approach-to-enhancing-manufacturing-productivity/#:~:text=Digital%20Twin%20Technology%20,time.

90. Dataparc. Digital Twins in Manufacturing Explained: Key Advantages and Implementation Insights [Available from: https://www.dataparc.com/blog/digital-twin-manufacturing/#:~:text=,allows%20engineers%20and%20operators%20to.

91. Shah V. Edge Computing and Digital Twins: A Match Made for Manufacturing [Available from: https://litmus.io/blog/edge-computing-and-digital-twins-a-match-made-for-manufacturing.

92. Shao G, Frechette SP, Srinivasan V, editors. An Analysis of the New ISO 23247 Series of Standards on Digital Twin Framework for Manufacturing2023 2023-06-16 04:06:00: 2023 MSEC Manufacturing Science & Engineering Conference, New Brunswick, NJ, US; 2023 MSEC Manufacturing Science & Engineering Conference, New Brunswick, NJ, US.

93. Anto Budiardjo DM. Digital Twin System Interoperability Framework

94. Park J. Manufacturing Execution System (MES) 2022 Guide for Batch Manufacturing [Available from: https://www.datacor.com/the-datacor-blog/manufacturing-execution-system-guide#:~:text=The%20diagram%20below%20illustrates%20where,making%20into%20the%20production%20operations.

95. Kakade S. manufacturing execution system (MES) [Available from: https://www.techtarget.com/searcherp/definition/manufacturing-execution-system-

MES#:~:text=A%20manufacturing%20execution%20system%20,opera tions%20and%20improve%20production%20output.

96. SAP. Manufacturing execution system overview [Available from: https://www.sap.com/products/scm/digital-manufacturing/what-is-mes.html#:~:text=ISA,logistics%20and%20process%20control%20syst ems.

97. Augello D. The Benefits of Integrating Your MES System With SCADA [Available from: https://inductiveautomation.com/blog/the-benefits-of-integrating-your-mes-system-with-scada#:~:text=Connecting%20layers%20of%20the%20manufacturing,l evels%20of%20the%20manufacturing%20process.

98. Stryker C. 12 considerations when choosing MES software [Available from: https://www.ibm.com/think/insights/manufacturing-execution-systems-software#:~:text=Acting%20as%20a%20connection%20between,raw% 20materials%20into%20finished%20products.

99. ibaset. MES and Industry 4.0 Manufacturing [Available from: https://www.ibaset.com/expertise/industry-4-0-mes/#:~:text=Closed%20Loop%20Integration.

100. EngIndX. 5 Ways Cloud MES Improves Your Production Capabilities [Available from: https://www.indx.com/en/posts/cloud-mes-improves-production#:~:text=A%20cloud%20MES%20is%20an,cloud%20MES% 20improves%20production%20by.

101. Shoplogix. MES in Food and Beverage: Transforming Production for Outstanding Results [Available from: https://shoplogix.com/mes-in-food-and-beverage/.

102. Zemsky J. PLM, ERP and MES: A Guide to Connected Systems in Discrete Manufacturing [Available from: https://www.ptc.com/en/blogs/plm/plm-erp-mes#:~:text=Chief%20among%20these%20systems%20are,individual %20roles%20of%20each%20system.

103. prescient. MES Implementation Challenges and How to Overcome Them [Available from: https://www.pre-scient.com/blogs/manufacturing-execution-systems/mes-implementation-challenges-and-how-to-overcome-them/#:~:text=Integration%20with%20Existing%20Systems.

104. SIEMENS. ISA-95 framework and layers [Available from: https://www.sw.siemens.com/en-US/technology/isa-95-framework-layers/.

105. 42Q. The Future of Automotive Manufacturing and MES [Available from: https://www.42-q.com/blog/news/the-future-of-

automotive-manufacturing-and-
mes/#:~:text=Manufacturing%20Execution%20Systems%20,traceabilit
y%20across%20the%20factory%20floor.
106. GE_Aerospace. GE Aerospace Improves Manufacturing
Productivity by Enabling a Smarter Factory [Available from:
https://www.gevernova.com/software/customer-stories/ge-aviation-
improves-manufacturing-productivity-with-smart-
factory#:~:text=GE%20Aerospace%20was%20operating%20six,and%
20drawings%20hard%20to%20manage.
107. BENG. Manufacturing Execution System Specialized for the
Pharmaceutical Industry: PAS-X MES Case Study [Available from:
https://www.b-en-g.co.jp/en/solution/case-
studies/case49.html#:~:text=%2A%20A%20GMP,pharmaceutical%20c
ompany%20they%20supply%20to.
108. Automation R. FactoryTalk PharmaSuite MES forms more
consistent high-quality foundation for ongoing digital transformation
[Available from: https://www.rockwellautomation.com/en-
us/company/news/case-studies/mes-forms-strong-foundation-for-
digital-
transformation.html#:~:text=MES%20forms%20strong%20foundation%
20for,deliver%20advanced%20data%20for.
109. Sepasoft. Driving Enterprise Efficiency and Operator
Engagement at Baldwin Richardson Foods [Available from:
https://www.sepasoft.com/case-studies/baldwin-richardson-oee-case-
study/.
110. Empowered. What is an Example of MES? [Available from:
https://www.empoweredautomation.com/what-is-an-example-of-
mes#:~:text=operations%2C%20helping%20to%20ensure%20efficienc
y%2C,high%20standards%20of%20product%20quality.
111. Gartner. Manufacturing Execution Systems Reviews and
Ratings [Available from:
https://www.gartner.com/reviews/market/manufacturing-execution-
systems.
112. contributors W. Wonderware [Available from:
https://en.wikipedia.org/w/index.php?title=Wonderware&oldid=1240468
154.
113. Schwarz M. AVEVA™ Manufacturing Execution System 2023
introduces an enterprise MES architecture [Available from:
https://www.aveva.com/en/perspectives/blog/aveva-manufacturing-
execution-system-2023-introduces-an-enterprise-mes-architecture-
that-helps-standardize-multi-site-
operations/#:~:text=AVEVA%E2%84%A2%20Manufacturing%20Execu

tion%20System%202023,of%20KPIs%2C%20reporting%20and%20sharing.

114.   Telefonica. MES systems for Industry 4.0 [Available from: https://www.telefonica.com/en/communication-room/blog/mes-systems-industry-4-0/.

115.   Eisner   C.   What   Is   ISA-95?   [Available   from: https://www.getmaintainx.com/learning-center/what-is-isa-95#:~:text=The%20ISA%20framework%20includes%20a,95%20framework.

116.   China CR. What is product lifecycle management (PLM)? [Available   from: https://www.ibm.com/think/topics/product-lifecycle-management#:~:text=What%20is%20product%20lifecycle%20management,PLM.

117.   Groisman J. Product lifecycle management [Available from: https://blogs.sw.siemens.com/teamcenter/what-is-plm-guide/.

118.   Chung E. The Role of PLM in Orchestrating the Product Digital Thread [Available from: https://www.spkaa.com/blog/the-role-of-plm-in-orchestrating-the-product-digital-thread#:~:text=PLM%20vs%20The%20Product%20Digital,Thread.

119.   ORACLE. What is PLM (Product Lifecycle Management)? [Available   from: https://www.oracle.com/emea/scm/product-lifecycle-management/what-is-plm/#:~:text=entire%20IT%20department%20to%20manage,costs%2C%20and%20enhanced%20product%20quality.

120.   Systemes D. PLM: Drive Innovation with Product Lifecycle Management   [Available   from: https://www.3ds.com/technologies/product-lifecycle-management.

121.   Arena.   Key   PLM   Capabilities   [Available   from: https://www.arenasolutions.com/what-is-plm/key-plm-capabilities/#:~:text=BOM%20Management.

122.   Grealou L. How configuration lifecycle management complements   PLM   and   MBSE   [Available   from: https://www.engineering.com/how-configuration-lifecycle-management-complements-plm-and-mbse/#:~:text=How%20configuration%20lifecycle%20management%20complements,variety%20and%20variability%20to.

123.   Oleg. A BOM Conflict Between PLM, ERP, and MES Systems [Available   from: https://beyondplm.com/2023/04/23/a-bom-conflict-between-plm-erp-and-mes-systems/#:~:text=A%20BOM%20Conflict%20Between%20PLM%2C,enabling%20process%20monitoring%20and%20optimization.

124. Groisman J. What is PLM? [Available from: https://blogs.sw.siemens.com/teamcenter/what-is-plm-guide/.
125. PTC. PLM (Product Lifecycle Management) [Available from: https://www.ptc.com/en/technologies/plm#:~:text=Volvo%20CE.
126. Windchill. Automotive PLM: Why You Need PLM for the Automotive Industry [Available from: https://nxrev.com/2024/10/automotive-plm/#:~:text=PLM%20enables%20teams%20to%20introduce,PLM%20powers%20a.
127. Autodesk. Data and process management that adapts to your business [Available from: https://www.autodesk.com/industry/design-manufacturing/plm/plm-product-lifecycle-management#:~:text=PLM%20enables%20real,and%20get%20to%20market%20faster.
128. LeverX. PLM Implementation Advice Every Business Owner Should Know About [Available from: https://leverx.com/newsroom/plm-implementation-advice-every-business-owner-should-know-about#:~:text=4.
129. Maniteja R. Transforming the Automotive Industry Through Product Lifecycle Management (PLM) [Available from: https://www.tek4s.com/post/transforming-the-automotive-industry-through-product-lifecycle-management-plm#:~:text=By%20leveraging%20PLM%2C%20automotive%20companies,can.
130. Ogewell V. Boeing, Airbus and the Hardship of Dealing with PLM Obsolescence [Available from: https://www.engineering.com/boeing-airbus-and-the-hardship-of-dealing-with-plm-obsolescence-tv-report/#:~:text=Obsolescence%20www,working%20flows%20between%20different%20systems.
131. Verma A. Streamline your Supply Chain Management with Autodesk's Fusion Manage for PLM [Available from: https://www.autodesk.com/blogs/design-and-manufacturing/2025/05/22/streamline-your-supply-chain-management-with-autodesks-fusion-manage-for-plm/.
132. Brohan M. Despite headwinds, more manufacturers will make AI a strategic priority [Available from: https://www.digitalcommerce360.com/2024/04/02/manufacturers-ai-strategic-priority/#:~:text=Table%20of%20Contents.
133. Schönfuß B. How AI is transforming the factory floor [Available from: https://www.weforum.org/stories/2024/10/ai-transforming-factory-floor-artificial-

intelligence/#:~:text=AI%20is%20rapidly%20transforming%20the,cost%20savings%20and%20reducing%20emissions.

134. AWS. What's the Difference Between AI and Machine Learning? [Available from: https://aws.amazon.com/compare/the-difference-between-artificial-intelligence-and-machine-learning/#:~:text=Artificial%20intelligence%20,AI%20activities%20are%20machine%20learning.

135. Wengel K. Working Smarter: How Manufacturers Are Using Artificial Intelligence [Available from: https://nam.org/wp-content/uploads/2024/05/NAM-AI-Whitepaper-2024-1.pdf#:~:text=%C2%84%20AI%20refers%20to%20a,improve%20work%20experience%20and%20create.

136. Delua J. Supervised versus unsupervised learning: What's the difference? [Available from: https://www.ibm.com/think/topics/supervised-vs-unsupervised-learning#:~:text=Within%C2%A0artificial%20intelligence%20%C2%A0,best%20approach%20for%20your%20situation.

137. Pal N. Reinforcement Learning in Practice: 10 Inspiring Examples to Learn From [Available from: https://emeritus.org/blog/best-reinforcement-learning-example/#:~:text=Manufacturing%20processes%20have%20been%20optimized,production%20efficiency%20and%20reduced%20waste.

138. EVIDEN. Computer Vision Reduces Quality Inspection Lead Time by 80% [Available from: https://eviden.com/insights/client-stories/computer-vision-helped-reduce-quality-inspection-time-in-manufacturing/#:~:text=Outcomes.

139. Lee S. Top 7 Ways Natural Language Processing Enhances Manufacturing [Available from: https://www.numberanalytics.com/blog/nlp-manufacturing-efficiency#:~:text=Maintenance%20represents%20one%20of%20manufacturing%E2%80%99s,predict%20failures%20before%20they%20occur.

140. Delfosse F. A Short Introduction to Generative Design [Available from: https://bigrep.com/posts/generative-design/#:~:text=manufacturer%20General%20Motors%20redesigned%20a,.

141. Akavia L. Artificial Intelligence And Machine Learning In Manufacturing: A Quick Guide To The Fundamentals [Available from: https://www.augury.com/blog/production-health/machine-learning-and-ai-in-manufacturing-a-quick-guide-to-the-fundamentals/#:~:text=Predictive%20Maintenance%C2%A0is%20the

%20more%20commonly,fairly%20common%20goal%20amongst%20 manufacturers.

142. Matthew Finio AD. How is AI being used in manufacturing? [Available from: https://www.ibm.com/think/topics/ai-in-manufacturing#:~:text=AI%20is%20also%20at%20the,more%20compl ex%20and%20creative%20work.

143. Peherstorfer A. AI and MES - Together into the future of manufacturing [Available from: https://www.industrieinformatik.com/en/newsbeitrag/ki-and-mes-together-into-the-future-of-manufacturing/#:~:text=manufacturing%20www,is%20revolutionizing% 20the%20industry.

144. JAGGAER. Manufacturing Execution Systems: How MES Transforms Production Efficiency [Available from: https://www.jaggaer.com/blog/manufacturing-execution-systems-mes#:~:text=Manufacturing%20Execution%20Systems%3A%20The% 20Future,business%20planning%2C%20inventory%20management%2 C.

145. Merkas W. The Impact of AI on Manufacturing Product Lifecycle Management [Available from: https://www.withum.com/resources/ai-manufacturing-product-lifecycle-management/#:~:text=1.

146. Columbus L. Optimizing Production Processes: The Role of IIOT & Connected Technology [Available from: https://blog.3ds.com/brands/delmia/optimizing-production-processes-the-role-of-iiot-connected-technology/.

147. Fujimaki R. The 6 Challenges of Implementing AI in Manufacturing [Available from: https://www.americanmachinist.com/enterprise-data/article/21149328/the-6-challenges-of-implementing-ai-in-manufacturing-dotdata.

148. Xie J, Sun L, Zhao YF. On the Data Quality and Imbalance in Machine Learning-based Design and Manufacturing—A Systematic Review. Engineering. 2025;45:105-31.

149. Zara D. AI Data Lifecycle Management: 6 Key Challenges [Available from: https://dialzara.com/blog/ai-data-lifecycle-management-6-key-challenges/.

150. Matellio. Implementing Edge Computing in Manufacturing: Enhancing Data Processing and Latency [Available from: https://www.matellio.com/blog/edge-computing-in-manufacturing/#:~:text=With%20traditional%20centralized%20cloud%2 0computing%2C,security%2C%20and%20overall%20system%20reliab ility.

151. Gao RX, Krüger J, Merklein M, Möhring H-C, Váncza J. Artificial Intelligence in manufacturing: State of the art, perspectives, and future directions. CIRP Annals. 2024;73(2):723-49.

152. Plathottam SJ, Rzonca A, Lakhnori R, Iloeje CO. A review of artificial intelligence applications in manufacturing operations. Journal of Advanced Manufacturing and Processing. 2023;5(3):e10159.

153. World I. What are the risks of using AI in manufacturing? [Available from: https://www.iiot-world.com/artificial-intelligence-ml/artificial-intelligence/what-are-the-risks-of-using-ai-in-manufacturing/#:~:text=Trust%20is%20paramount%20when%20it,way%20that%20builds%20similar%20trust.

154. CENTRIPETAL. How Does IIOT Impact Cybersecurity In Manufacturing? [Available from: https://www.centripetal.ai/blog/what-is-iiot-and-how-does-it-impact-cybersecurity-in-manufacturing/#:~:text=How%20Does%20IIOT%20Impact%20Cybersecurity,exploit%20the%20larger%20attack.

155. Bots S. Lights-out manufacturing: Full automation on the horizon [Available from: https://standardbots.com/blog/lights-out-manufacturing?srsltid=AfmBOop4WTbcqmZ-ckCsCECiYG6WKJ7z_LCCFhvkkbPvwQo4eg8scwjk#:~:text=Lights,Once%20systems.

156. Islam F, Raihan A, Ahmed I. Applications of Federated Learning in Manufacturing: Identifying the Challenges and Exploring the Future Directions with Industry 4.0 and 5.0 Visions2023.

157. Stryker C. Artificial intelligence implementation: 8 steps for success [Available from: https://www.ibm.com/think/insights/artificial-intelligence-implementation#:~:text=Step%201%3A%20Define%20goals.

158. Hyland S. AI integration in Manufacturing: Best Practices and Use Cases [Available from: https://veriday.com/ai-integration-in-manufacturing-best-practices-and-use-cases/#:~:text=1,to%20ensure%20that%20your%20AI.

www.ingramcontent.com/pod-product-compliance
Lightning Source LLC
Chambersburg PA
CBHW030717250326
R18027900001B/R180279PG41599CBX00020B/31

* 9 7 8 1 7 6 3 7 7 7 8 8 0 *